Automated Design of Electrical Converters with Advanced AI Algorithms

Xin Zhang · Xinze Li · Hao Ma · Jingjing Huang · Zeng Zheng

Automated Design of Electrical Converters with Advanced AI Algorithms

Springer

Xin Zhang
College of Electrical Engineering
Zhejiang University
Hangzhou, Zhejiang, China

Hao Ma
College of Electrical Engineering
Zhejiang University
Hangzhou, Zhejiang, China

Zeng Zheng
Chongqing University
Chongqing, China

Xinze Li
Nanyang Technological University
Singapore, Singapore

Jingjing Huang
Department of Electrical Engineering
Xi'an University of Technology
Xi'an, Shaanxi, China

ISBN 978-981-99-0458-7 ISBN 978-981-99-0459-4 (eBook)
https://doi.org/10.1007/978-981-99-0459-4

This Springer imprint is published by the registered company Springer Nature Singapore Pte Ltd.
The registered company address is: 152 Beach Road, #21-01/04 Gateway East, Singapore 189721, Singapore

Preface

In power engineering, electrical engineering, and electric power industry, power converters play an important role in converting electricity from one form to another such as converting between AC and DC; or regulating the voltage or frequency; or some combinations of these. It is required for a variety of applications, such as industrial drives, power supplies, energy production equipment, consumer goods, electric vehicles/aircrafts/ships, and smart grid. Because the applications of power converters become more and more complex and diverse, its design requirements become increasingly difficult for engineers. For example, if a power converter is designed, its design target may include many optimization objectives, such as efficiency, power density, control performance, system stability, reliability, and so on. It is not possible to require human beings to find the optimal design only by personal experience. Consequently, how to design power converters in a reasonable way to meet the above complex and diverse requirements becomes an urgent concern of the engineers.

This book will open a new door for the world to design the power converters through artificial intelligence (AI) tools. This book first reviews the existing AI technologies in power converters. Furthermore, this book introduces special AI algorithms for power converters, which considers the unique characteristics of power converters. After that, this book introduces a family of our proposed AI-based design method for the power devices, DC/DC converters, resonant DC/DC converters, bidirectional DC/DC converters, and DC/AC inverters. To the best of the authors' knowledge, this book is the first book that comprehensively introduces how to use the AI technologies in the design of power converters, including literature review, algorithm, circuit design, control strategy, etc.

This book comprehensively provides self-contained knowledge regarding the AI-based design of power converters from Prof. Zhang Xin's research Group. The purposes of this book are to introduce the AI-based design methodologies for the power converters in a systematical way, so as to reflect the superiority of the AI-based design methods and eventually attract more people to adopt these useful methods.

This book is suitable for the following disciplines: power electronics, control engineering, artificial intelligent, and circuit design. This book can be treated as the graduate textbooks, undergraduate textbooks, and industrial design instruction books of the above areas, which can bring benefits to both academia and industry.

Hangzhou, China Xin Zhang

Contents

About the Authors

Xin Zhang received the Ph.D. degree in automatic control and systems engineering from the University of Sheffield, UK, in 2016 and the Ph.D. degree in electronic and electrical engineering from Nanjing University of Aeronautics and Astronautics, China, in 2014. From February 2014 to December 2016, he was Research Associate at the University of Sheffield. From January 2017 to September 2017, he was Postdoctoral Research Fellow at the City University of Hong Kong. From September 2017 to August 2020, he was Assistant Professor of power engineering at the School of Electrical and Electronic Engineering, Nanyang Technological University, Singapore. Currently, he is Full Professor at Zhejiang University. He is generally interested in power electronics, power systems, and advanced control theory, together with their applications in various sectors. He has received the Highly Prestigious Chinese National Award for Outstanding Students Aboard in 2016. He is Associate Editor of 8 SCI journals and transactions, such as IEEE TIE/JESTPE/OJPE Access and IET Power Electronics.

Mr. Xinze Li received bachelor's degree in power engineering from Shandong University, China, in 2018, and currently working toward the Ph.D. degree in electrical and electronic engineering at Nanyang Technological University, Singapore. He is the IEEE student member since 2018. His research interests include the applications of artificial intelligence in power electronics, fully automated design of power converters, DC-DC converter, circuit parameter design, modulation design, filter design, electric vehicle, renewable energy, deep learning, computer vision, natural language processing, and evolutionary algorithm.

Prof. Hao Ma received the B.S., M.S., and Ph.D. degrees from Zhejiang University, Hangzhou, China, in 1991, 1994, and 1997, respectively, all in Electrical Engineering. Since 1997, he worked as a lecturer, associate professor, and professor at Zhejiang University. Currently, he is a professor in the College of Electrical Engineering and served as Vice Dean of ZJU-UIUC Institute, Zhejiang University. His research interests include advanced control in power electronics, wireless power transfer, fault diagnosis of power electronic circuits and systems, and application of power electronics.

He is served as Director of Academic Committee of China Power Supply Society, Vice President, and Secretary-general of Power Supply Society of Zhejiang Province, Associate Editor of *Journal of Power Electronics* (JPE), Associate Editor of *IEEE Journal of Emerging and Selected Topics in Power Electronics* (JESTPE). He was served as Vice Dean of College of Electrical Engineering, Zhejiang University (2013–2017), AdCom member of IEEE Industrial Electronics Society (2014–2015), Technical Program Chair of IEEE PEAC 2018 and IEEE PEAC 2014, Technical Program Chair of IEEE ISIE 2012 and was served as Co-chair of Power Systems and Smart Grids, IEEE IECON 2018, Special Session Co-chair of IEEE IECON 2017, Co-chair of Power Electronics and Energy Conversion, IEEE IECON 2013, and Co-chair of Power Electronics and Renewable Energy Track, IEEE IECON 2010.

Professor Jingjing Huang received the B.S. degree from the Henan University of Science and Technology, Luoyang, China, in 2008, and the Ph.D. degree from Xi'an Jiaotong University, Xi'an, China, in 2014, both in electrical engineering.

From April 2014 to November 2016, she was full-time Lecturer at the Xi'an University of Technology, Xi'an. From December 2016 to December 2019, she was full-time Postdoctoral Research Fellow at Nanyang Technological University, Singapore. She is currently Associate Professor at the School of Electronic and Information Engineering, Xi'an Jiaotong University. Her research interests include renewable energy systems, high-frequency transformer, hybrid AC/DC microgrid, and high-power converters.

Professor Zeng Zheng received the B.Sc. degree in electrical engineering from Wuhan University, Wuhan, China, in 2009, and the Ph.D. degree in electrical engineering from Zhejiang University, Hangzhou, China, in 2014. He joined School of Electrical Engineering, Chongqing University, Chongqing, China, in July 2014, where he was promoted to associate professor in August 2017. From July 2018 to July 2019, he was a research fellow with School of Electrical and Electronic Engineering, Nanyang Technological University, Singapore. His research interests include grid-connected inverter for renewable energy integration and customized package for wideband gap power devices.

Chapter 1
Introduction

1.1 Backgrounds

1.1.1 Basic Introduction to DC-DC Converters

In the current era, environmental problems such as the global warming and the depletion of fossil fuel attract more and more attentions globally, leading to the increasing penetration of renewable energy resources [1, 2]. The inherently fluctuating, intermittent and widely-distributed renewable energy brings challenges to the existing power grid. To ensure the reliable and stable connections between renewable energy resources and existing power grid, DC-DC converters and DC-AC inverters are the key enablers. Several applications of DC-DC converters and DC-AC inverters are shown in Fig. 1.1.

Generally, the topologies of DC-DC converters can be classified into non-isolated and isolated types according to whether there is galvanic isolation, or can be classified into unidirectional and bidirectional types based on the directions of power transfer [3]. In the non-isolated DC-DC converters, conventional Buck, Boost, Buck-Boost, Ćuk and Sepic/Zeta converters have been widely applied in photovoltaic systems [4], LED drivers [5], fuel cell vehicles [6], etc. To enhance the voltage boost ability and relieve the burden of current stress, two or more converters can be connected in a cascaded fashion, which has applications such as smart grids and distributed power systems [2]. Interleaved and multilevel converters, which are also non-isolated DC-DC converters, can be utilized in automotive systems [7], HVDC grids [8], etc.

From the perspective of isolated DC-DC converters, except for the benefits of safety, the galvanic isolation can achieve high voltage gain ratio and provide the possibility of multi-input and multi-output topologies. In the isolated DC-DC converters, flyback, push–pull and forward topologies are commonly applied in low and medium power situations [3]. The most popular isolated DC-DC converters are dual active bridge (DAB) converters, whose applicational fields include electric vehicle charging

X. Zhang et al., *Automated Design of Electrical Converters with Advanced AI Algorithms*,
https://doi.org/10.1007/978-981-99-0459-4_1

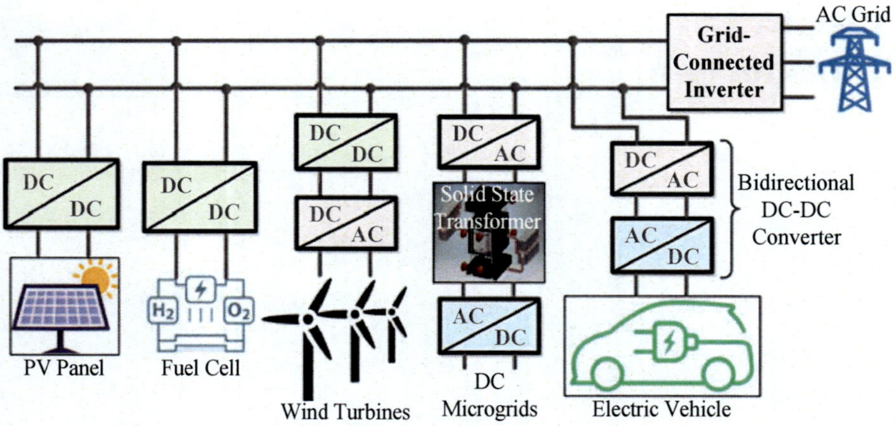

Fig. 1.1 Applications of DC-DC converters and DC-AC inverters

[7], battery storage systems [9], uninterrupted power supply [3], solid-state transformer [10], etc. Recently, DAB converters with resonant tanks such as LC, LLC, CLLC [11] are hot research topics.

1.1.2 Basic Introduction to DC-AC Inverters

To build the connection between the existing AC power grids and the distributed renewable energy sources such as solar energy, wind energy and tidal energy, DC-AC inverters are also playing significant roles, as the applications shown in Fig. 1.1. According to the operation modes, inverters can be classified into stand-alone, grid-connected and bimodal types [12, 13]. From the configuration topology point of view, the grid-connected inverters mainly consist of three types: centralized inverters, string inverters and module inverters, which are suitable for high power, medium power and low power applications [14, 15]. Similar to DC-DC converters, DC-AC inverters are widely used in renewable energy systems, such as solar PV systems [16] and wind farm systems [17, 18]. In addition, Inverters can be used in space applications [19], telecommunication [20], computer systems [21] and traction scenarios [22].

1.2 Problem Descriptions

Due to the increasing proportion of DC-DC converters and DC-AC inverters in both industry and our daily life, their power modules, circuit parameters and modulation schemes have to be carefully studied and optimized from power quality, stability, efficiency, reliability and economic perspectives [16, 23, 24]. In practice, to design and deploy a power electronic system in real-world, the whole process mainly consists of three steps: performance analysis, optimization, and modulation and control.

As exhibits in Fig. 1.2, the conventional methods for performance analysis, optimization and modulation of power converters suffer from some unneglectable problems: time-consuming and inaccurate manual performance analysis, suboptimal optimization and unsatisfactory modulation [25, 26]. Generally, conventional performance analysis suffers from excessive human-dependency, time-consuming and inaccurate problems, traditional optimization approaches suffer from suboptimal problem and heavy computation, and conventional modulation approaches may result in high space and time complexity.

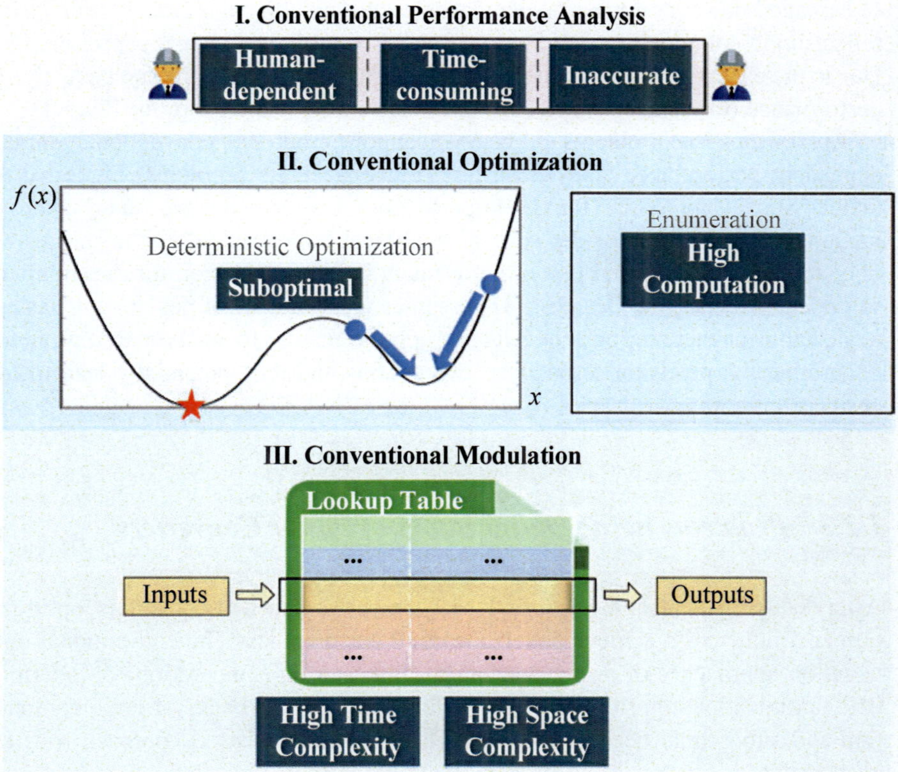

Fig. 1.2 Problems in the conventional approaches for the performance analysis, optimization and modulation for DC-DC converters and DC-AC inverters

1.2.1 Problems in the Human-Dependent Analysis of Performance of Power Converters

In the performance analysis of power converters, most of the existing conventional methods require the fully involvements of engineers and researchers for the mathematical expressions. For example, to obtain the expressions of copper loss and conduction loss of DAB converter, the inductor current has to be analyzed piece by piece for all the operating modes, and the deduced high-order expressions are then squared and integrated to compute the rms value for further computing copper and conduction losses [27]. Lin et al. manually deduces the analytical expressions of the output impedance of symmetrical CLLC-type DAB converter for the stable operation of cascaded power converter systems, where the small-signal model of main circuit reflects high complexity caused by the four resonant inductors and capacitors [28]. Due to the excessive human-dependency, the conventional analysis and deduction of performance of power converters are time-consuming and error prone [29, 30].

Apart from the problems of heavy manpower burden, conventional human-dependent performance analysis suffers from inaccurate problems. To derive the mathematical expressions for the targeted optimization objective, many approximations will be taken for the sake of analytical convenience. For instance, [23] only adopts the zeroth and first order terms in Fourier expansion for the simplicity of impedance analysis. Besides, [31] assumes ideal models of power switches and neglectable magnetizing current. All these approximations for analytical convenience undermine the precision of deduced expressions and thus potentially lead to low design accuracy.

1.2.2 Problems in the Optimization of Power Converters

After deducing the analytical formulas for the performance of converters, optimization is conducted to achieve better operating performance. The conventional optimization approaches of power converters are based on enumerations or deterministic optimization algorithms, which suffer from the drawbacks of large computation and suboptimal designs [32, 33]. The enumeration-based approach tries out all possible design cases, so its computational burden for computer is extremely heavy, while the design cycle is an important factor to be considered. As shown in Fig. 1.2 [34], enumeration-based approaches for the optimization of DC-DC converters may require infeasible computational time especially when the dimension of design parameters is high or when the design parameters are continuous with infinite possible values.

As for the deterministic optimization approaches such as sequential unconstrained minimization, augmented Lagrangian and Newton–Raphson, their optimization results highly depend on the selected initial iteration points and the predetermined convergence threshold, as plotted in the left-side of Fig. 1.2. Both improper initial iteration point and inadequate convergence threshold will lead to unsatisfactory local optima [35]. With deterministic approaches, the optimality of design may be nontrivially undermined with a high possibility. Busquets-Monge et al. [36] reveals the suboptimal problems of some commonly used deterministic algorithms in the optimization of boost power factor correction converter.

1.2.3 Problems in the Modulation of Power Converters

To realize the real-time modulation for power converters, there are mainly two approaches: through online computation of deduced expressions or through searching a stored lookup table. For instance, [37] adopts lookup table to realize online triple phase shift modulation of DAB converter, in which the implemented lookup table stores the optimal modulation parameters for different operating situations. Zhou and Wang [38] utilizes deduced formulas to realize real-time space vector modulation (SVM).

Generally, the space and time complexity of the real-time modulation are always considered, both of which are favoured to be as low as possible. It has to be admitted that the conventional lookup-table-based and expression-based modulation schemes are easy to implement, but they suffer from either high space complexity or unacceptable time complexity. According to Tang [37], the storage size of lookup table exponentially rises with the increasing of stored variables, and the exploding storage size will nontrivially slow down the query speed of lookup table, as shown in the bottom of Fig. 1.2. Compared to lookup tables, expression-based modulation approaches have low space complexity, but the time complexity might be unacceptable due to the non-linearity of the expressions required to be solved in real-time [39]. Apart from that, the deduction process for the expression-based modulation approaches also suffers from tedious and overwhelming human-dependency.

1.3 Basic Introduction to Artificial Intelligence Algorithms

To overcome the aforementioned problems in the analysis, optimization and modulation of power converters, three AI tools can be applied: neural networks, evolutionary algorithms and fuzzy inference systems, which are introduced one by one as the followings.

1.3.1 Neural Networks

Neural network (NN), which lies at the heart of recent popular deep learning algorithms, is a parallel-structured computational graph mimicking the flexible connections of biological neurons in brain [40]. As shown in Fig. 1.3, the general structure of NN consists of input layer, hidden layers and output layer, which are responsible for obtaining inputs, learning underlying behaviors and predicting outputs. Artificial neurons in NN connect with one another through adjustable weights and biases and non-linear activation functions. Being beneficial from its high non-linearity and easily adjustable structure, NN can learn any complex and nonlinear relationships between design variables and objectives with arbitrary precision, so it has better fitting accuracy compared with other regression techniques such as ridge regression, Bayesian regression and support vector regression [41].

In the current literatures, three types of NN are commonly adopted, namely feed-forward NN (FNN), recurrent NN (RNN) and convolutional NN (CNN) [42], the standard structures of which are shown in Fig. 1.3. FNN passes the information in only one direction, from inputs to hidden layers and then to outputs. CNN utilizes convolutional and pooling layers to handle image-like data, so CNN is widely used in image recognition, object detection, and video detection [43]. Different from FNN which does not have cycles or loops in the network, RNN utilizes loops to recurrently feed the historical information to the current output, exhibiting temporal dynamic behaviors. RNN is proficient in handling time-series data, so is widely used in time series prediction, natural language processing, voice detection, and control [44].

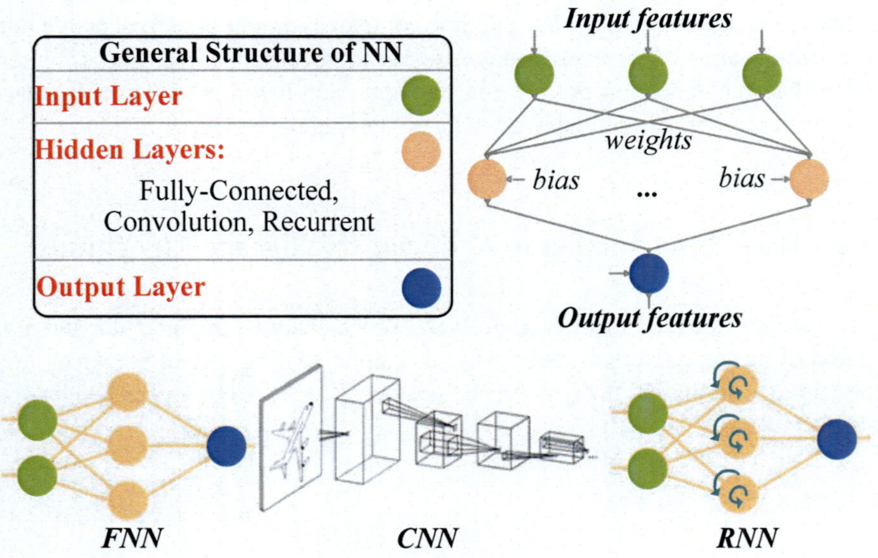

Fig. 1.3 General structure of NN and three types of commonly seen networks (FNN, CNN and RNN)

1.3.2 Evolutionary Algorithms

Evolutionary algorithms (EAs) are stochastic optimization techniques, which are derived from biological phenomenon or physical processes [45]. For example, genetic algorithm (GA) imitates the environmental selection in nature [46]. Particle swarm optimization (PSO) is derived from the social behaviors of bird flocks [47]. Ant colony optimization (ACO) simulates the information tracking behavior of ants to find the optimal solutions [48]. The key characteristics of EAs are that better individuals have larger chances of survival and reproduction, while worse individuals can still survive and reproduce, because of which the solutions can maintain good diversity while still can converge to optimum. Among the three EAs, PSO algorithm is suitable for continuous optimization, ACO specializes in solving discrete optimization problems, and GA performs the best in mixed-integer optimization [45].

According to the number of design objectives to be optimized, EAs are divided into single-objective and multi-objective algorithms [49]. The common steps of EAs shown in Fig. 1.4 are discussed as follows. Firstly, the hyperparameters and states of algorithms are initialized, and the fitness or objective functions of individuals are then evaluated. Subsequently, individuals will be selected to generate new solutions according to the values of fitness or objective functions, and better ones stand larger chance of survival. Afterwards, new solutions are compared to original ones and get updated. This process repeats until the stopping criterions have been met.

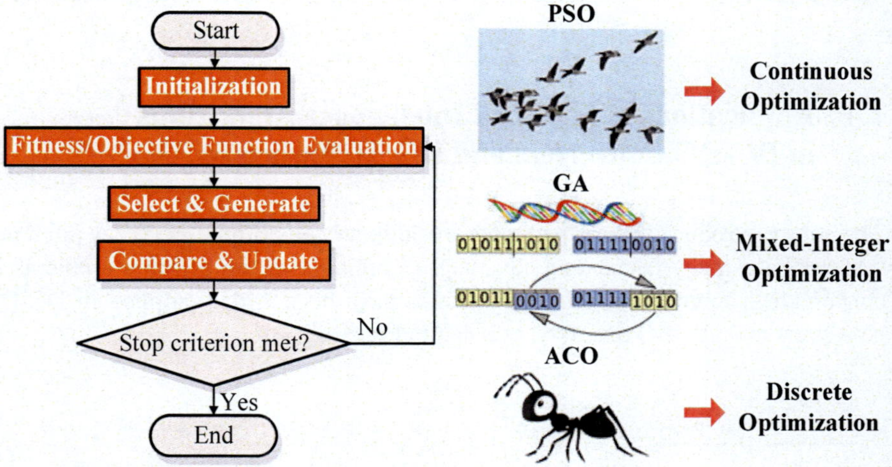

Fig. 1.4 Common process of EA and three popular algorithms (PSO, GA, ACO)

1.3.3 *Fuzzy Inference System*

Taking advantage of fuzzy set theory, fuzzy inference system (FIS) can properly interpret the ambiguity and fuzziness of real-world data. FIS transfers the usual crisp value (0 or 1) to a degree of truth (continuous between 0 and 1), which provides human-like logic that enables intermediate degrees other than true or false [50, 51]. According to the type of its output function, FIS is classified into Mamdani and Sugeno types [52], which are shown in Fig. 1.5.

The output function of Mamdani FIS belongs to fuzzy membership functions, and a typical Mamdani FIS contains four steps: fuzzification, fuzzy inference, fuzzy rules congregation and defuzzification. Fuzzification step computes the membership degree of input variables belonging to each fuzzy linguistic set. Fuzzy inference evaluates fuzzy rules. The results of all fuzzy rules are congregated and defuzzified to calculate the outputs. Mamdani FIS can be implemented in both multiple input and single output (MISO) system and multiple input and multiple output (MIMO) system [52]. Compared to Mamdani FIS whose outputs are fuzzy membership functions, Sugeno FIS has no fuzzy membership in its output, and it only utilizes crisp functions as the output functions. Sugeno FIS computes its final output through the mathematical combinations of crisp values and rule firing strength, which is different from the defuzzification step in Mamdani FIS. Sugeno FIS possesses more flexibility than Mamdani FIS, but it can only be used MISO system. If the fuzzy membership functions and rules of FIS are properly tuned, the whole system will successfully track desired behaviors (regression) or correctly classify the data (classification).

1.4 Applications of Artificial Intelligence Algorithms in DC-DC Converters and DC-AC Inverters

The briefly introduced neural networks, evolutionary algorithms and fuzzy inference systems have many fascinating advantages in mitigating the possible problems of conventional approaches. These algorithms have been widely applied in DC-DC converters and DC-AC inverters, as listed in Fig. 1.6.

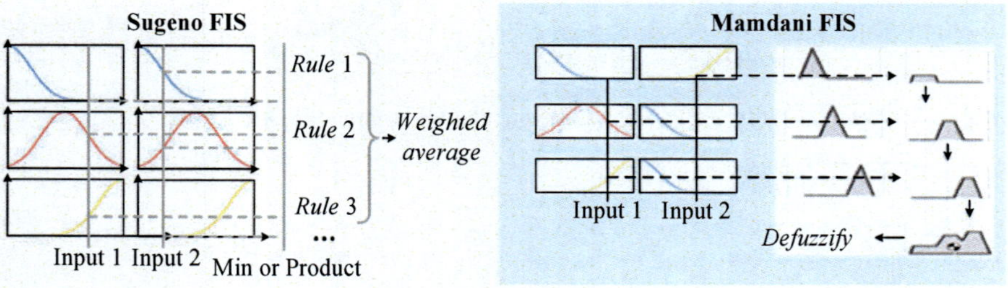

Fig. 1.5 Sugeno FIS and Mamdani FIS

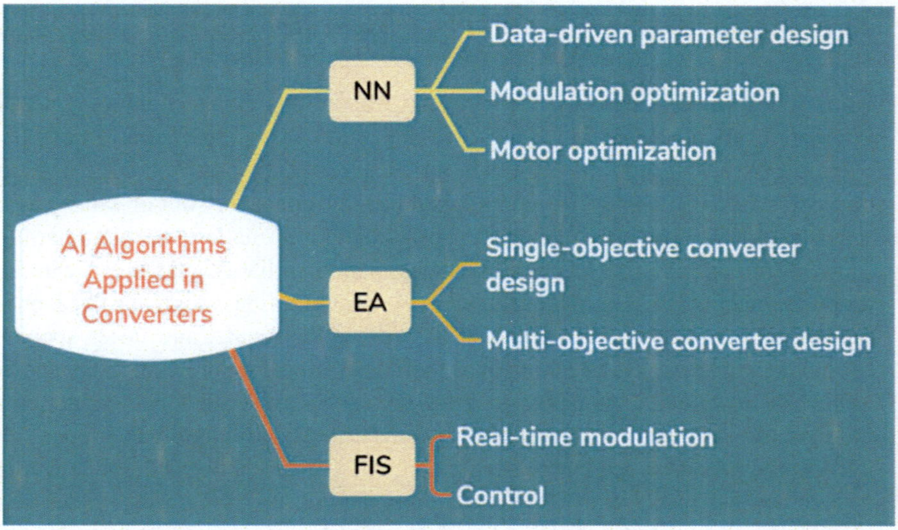

Fig. 1.6 Applications of NN, EA and FIS in DC-DC converters and DC-AC inverters

1.4.1 Applications of Neural Networks

NN has strong capability in memorization and generalization attributable to the adjustable weights, non-linear activation functions and extendible network structures. If sufficient performance data such as efficiency, current stress and reliability is provided, NN can be trained to automatically learn the underlying mathematical expressions, serving as accurate data-driven models to substitute time-consuming human-dependent performance analysis [53, 54]. Utilizing NN, the drawbacks of heavy manpower burden and inaccurate performance analysis in conventional human-dependent approaches can be largely relieved.

NN has been widely adopted in parameter design, modulation strategy optimization and motor optimization [53, 55, 56]. For instance, in [53], neural network (NN) is trained with simulation results to act as the surrogate model for component lifetime consumption for reliability-oriented parameter design. Moreover, in [55], extreme learning machine is adopted to replace traditional model deduction of ripple, harmonics and transient response of permanent magnetic synchronous linear motors to achieve optimal design of permanent magnetic motor. Kazmierkowski et al. [56] utilizes NN to realize adaptive selective harmonic elimination for cascaded multilevel inverters with varying DC sources. NN serves as the model predictive controller for modular multilevel DC-DC converters in [57]. In [58] with regards to SVM for voltage-fed inverter induction motor drive, neural network learns the switching time of reference vectors, reduces the required online computation and raises the frequency limit. In the fault diagnosis of multilevel modular converter, a convolutional neural network is used and achieves high accuracy, with a good noise tolerant ability [59].

1.4.2 Applications of Evolutionary Algorithms

From the descriptions in Sect. 1.2.2, conventional optimization techniques generally suffer from suboptimal design results and high computation problems. To alleviate the problems of conventional optimization techniques, EAs can be used. EAs are stochastic and meta-heuristic searching methods, the characteristics of which ensure their searching performance independent of gradient information and initial iteration points. With EAs, the globally optimal design can be reliably guaranteed with low computational burden [45]. As discussed in the configuration design of a three-phase wound core transformer [33], the better optimization results and faster speed of GA and differential evolution compared with deterministic optimization are shown. Another advantage of EAs is that they require no rigid mathematical deduction as deterministic optimizations, and the algorithms are easy for implementation.

Recently, EAs have been increasingly applied in the design of DC-DC converters and inverters. For instance, in [34], simulated annealing algorithm has been adopted to optimize the parameter values of DC-DC converter. Considering the infinite possible combinations of varying parameter values, PSO algorithm is adopted in [10] to achieve stability, acceptable efficiency and robust voltage conversion gain of DC-DC converter cascading with a Buck converter. In the series–parallel RLC filter automatic synthesis, GA is applied for the optimal filter topology design [60]. If multiple optimization objectives are considered simultaneously, the multi-objective evolutionary algorithms can be used. For example, the efficiency, reliability and cost of distributed maximum power point tracking converter [61] are optimized by non-dominated sorting genetic algorithm—II (NSGA-II). With the output LC filter as design parameters [62], holistic performance of DC-DC converter is realized via NSGA-II, with the simultaneous considerations of reliability, volume, and cut-off frequency. Furthermore, by tuning the frequency, current density, magnetic influx, transformer topology and material, and type of power switches, the weight, cost and power loss of isolated DC-DC converter are minimized altogether through NSGA-II [42].

1.4.3 Applications of Fuzzy Inference Systems

To solve the problems of high space and time complexity of conventional lookup-table-based approaches and formula-based approaches for real-time modulation, FIS can be utilized. Compared with the conventional online modulation approaches, the time and space complexity of FIS is independent of data size, and a carefully-tuned FIS always requires low storage size and has fast computation speed [63, 64]. Apart from its superior algorithm complexity, FIS has other appealing advantages such as easy implementation, good linguistic interpretability, and satisfactory generalization capability [51].

Being beneficial from all of its merits, FIS has been widely implemented in the real-time modulation for power converters. For instance, in the maximum power point tracking of the boost-converter-supplied PV system [8], the fuzzy inference controller, substituting the conventional PID controllers, manifests strong robustness against the fluctuations of parameters, loads, and supply voltage. In terms of SVM for a two-level inverter [65], adaptive network-based FIS has been used, which has realized smaller harmonic distortions than conventional formula-based SVM. Li has proposed a FIS-based modulation scheme for current-stress-minimized triple phase shift modulation for isolated DAB converter [66] under varying operating power and voltage. In [67], the network-based FIS approach offers extremely fast dynamic response with high accuracy, and effectively controls the injected power and maintains the stringent voltage, current, and frequency conditions. Saroha et al. [68] presents an adaptive network-based FIS controller for the unbalanced voltage compensation in a low-voltage microgrid with multiple voltage source converters.

1.5 Arrangement of This Book

Being inspired by the outstanding performance of AI algorithms, this book aims at exploring the applications of AI algorithms in the design and modulation of DC-DC converters and DC-AC inverters. The rest of this book is organized as follows. In Chap. 2, to overcome the under-optimization problem and promote the performance of the emerging double-sided cooling power module, a multi-objective design methodology considering thermal and mechanical performance is proposed. In Chap. 3, a novel coevolving archived-based multi-objective algorithm is proposed, based on which the output LC filter in Buck converter can be flexibly designed to meet requirements in various applications while maintaining outstanding comprehensive performance. Chapter 4 introduces an artificial-intelligence-based design approach for the circuit parameters of power converters, which utilizes the integration of simulation, NN and EA to realize a high-level automation in performance analysis and optimization. Aiming at the optimization of the total power loss in the LCLC resonant converter for the space travelling-wave tube amplifier applications, an efficiency-oriented two-stage optimal design methodology has been proposed. In Chap. 5, for hybrid AC/DC microgrid applications, an AI-based (GA + PSO) two-stage optimal design methodology for high-efficiency CLLC resonant converters has been proposed. Chapter 6 proposed an AI system to minimize the current stress of DAB converter under triple phase shift modulation, in which simulation and NN are integrated to realize data-driven current stress analysis, PSO algorithm is used to search for optimal modulation parameters, and FIS is applied online for satisfactory real-time modulation. In Chap. 7, a systemic co-design methodology is proposed for the overall optimization of the air-cooling SiC inverter from the perspectives of thermal, electrical and mechanical performance. In the end, the conclusion is summarized in Chap. 8.

References

1. J. M. Carrasco *et al.*, 'Power-Electronic Systems for the Grid Integration of Renewable Energy Sources: A Survey', *IEEE Trans. Ind. Electron.*, vol. 53, no. 4, pp. 1002–1016, Jun. 2006.https://doi.org/10.1109/TIE.2006.878356
2. F. Blaabjerg, R. Teodorescu, M. Liserre, and A. V. Timbus, 'Overview of Control and Grid Synchronization for Distributed Power Generation Systems', *IEEE Trans. Ind. Electron.*, vol. 53, no. 5, pp. 1398–1409, Oct. 2006.https://doi.org/10.1109/TIE.2006.881997
3. S. A. Gorji, H. G. Sahebi, M. Ektesabi, and A. B. Rad, 'Topologies and Control Schemes of Bidirectional DC–DC Power Converters: An Overview', *IEEE Access*, vol. 7, pp. 117997–118019, 2019.https://doi.org/10.1109/ACCESS.2019.2937239
4. W. Li and X. He, 'Review of Nonisolated High-Step-Up DC/DC Converters in Photovoltaic Grid-Connected Applications', *IEEE Trans. Ind. Electron.*, vol. 58, no. 4, pp. 1239–1250, Apr. 2011.https://doi.org/10.1109/TIE.2010.2049715
5. Y. Wang, J. M. Alonso, and X. Ruan, 'A Review of LED Drivers and Related Technologies', *IEEE Trans. Ind. Electron.*, vol. 64, no. 7, pp. 5754–5765, Jul. 2017.https://doi.org/10.1109/TIE.2017.2677335
6. M. Sagar Bhaskar *et al.*, 'Survey of DC-DC Non-Isolated Topologies for Unidirectional Power Flow in Fuel Cell Vehicles', *IEEE Access*, vol. 8, pp. 178130–178166, 2020. https://doi.org/10.1109/ACCESS.2020.3027041
7. A. Khaligh and S. Dusmez, 'Comprehensive Topological Analysis of Conductive and Inductive Charging Solutions for Plug-In Electric Vehicles', *IEEE Trans. Veh. Technol.*, vol. 61, no. 8, pp. 3475–3489, Oct. 2012.https://doi.org/10.1109/TVT.2012.2213104
8. J. D. Paez, D. Frey, J. Maneiro, S. Bacha, and P. Dworakowski, 'Overview of DC–DC Converters Dedicated to HVdc Grids', *IEEE Trans. Power Delivery*, vol. 34, no. 1, pp. 119–128, Feb. 2019.https://doi.org/10.1109/TPWRD.2018.2846408
9. G. Wang *et al.*, 'A Review of Power Electronics for Grid Connection of Utility-Scale Battery Energy Storage Systems', *IEEE Trans. Sustain. Energy*, vol. 7, no. 4, pp. 1778–1790, Oct. 2016.https://doi.org/10.1109/TSTE.2016.2586941
10. F. Lin, X. Zhang, and X. Li, 'Design Methodology for Symmetric CLLC Resonant DC Transformer Considering Voltage Conversion Ratio, System Stability, and Efficiency', *IEEE Trans. Power Electron.*, vol. 36, no. 9, pp. 10157–10170, Sep. 2021.https://doi.org/10.1109/TPEL.2021.3059852
11. B. Zhao, Q. Song, W. Liu, and Y. Sun, 'Overview of Dual-Active-Bridge Isolated Bidirectional DC–DC Converter for High-Frequency-Link Power-Conversion System', *IEEE Trans. Power Electron.*, vol. 29, no. 8, pp. 4091–4106, Aug. 2014.https://doi.org/10.1109/TPEL.2013.2289913
12. A. Kumar and P. Sensarma, 'A Four-Switch Single-Stage Single-Phase Buck–Boost Inverter', *IEEE Trans. Power Electron.*, vol. 32, no. 7, pp. 5282–5292, Jul. 2017.https://doi.org/10.1109/TPEL.2016.2605150
13. M. N. H. Khan, M. Forouzesh, Y. P. Siwakoti, L. Li, T. Kerekes, and F. Blaabjerg, 'Transformerless Inverter Topologies for Single-Phase Photovoltaic Systems: A Comparative Review', *IEEE J. Emerg. Sel. Topics Power Electron.*, vol. 8, no. 1, pp. 805–835, Mar. 2020.https://doi.org/10.1109/JESTPE.2019.2908672
14. M. Calais, J. Myrzik, T. Spooner, and V. G. Agelidis, 'Inverters for single-phase grid connected photovoltaic systems-an overview', in *2002 IEEE 33rd Annual IEEE Power Electronics Specialists Conference. Proceedings (Cat. No.02CH37289)*, Cairns, Qld., Australia, 2002, vol. 4, pp. 1995–2000. doi: https://doi.org/10.1109/PSEC.2002.1023107.
15. S. B. Kjaer, J. K. Pedersen, and F. Blaabjerg, 'A Review of Single-Phase Grid-Connected Inverters for Photovoltaic Modules', *IEEE Trans. on Ind. Applicat.*, vol. 41, no. 5, pp. 1292–1306, Sep. 2005.https://doi.org/10.1109/TIA.2005.853371
16. Q. Huang, A. Q. Huang, R. Yu, P. Liu, and W. Yu, 'High-Efficiency and High-Density Single-Phase Dual-Mode Cascaded Buck–Boost Multilevel Transformerless PV Inverter With GaN

AC Switches', *IEEE Trans. Power Electron.*, vol. 34, no. 8, pp. 7474–7488, Aug. 2019.https://doi.org/10.1109/TPEL.2018.2878586

17. Arani and El-Saadany, 'Implementing Virtual Inertia in DFIG-Based Wind Power Generation', *IEEE Trans. Power Syst.*, vol. 28, no. 2, pp. 1373–1384, May 2013.https://doi.org/10.1109/TPWRS.2012.2207972
18. M. Liserre, R. Teodorescu, and F. Blaabjerg, 'Stability of photovoltaic and wind turbine grid-connected inverters for a large set of grid impedance values', *IEEE Trans. Power Electron.*, vol. 21, no. 1, pp. 263–272, Jan. 2006.https://doi.org/10.1109/TPEL.2005.861185
19. P. Jain, M. Pahlevaninezhad, S. Pan, and J. Drobnik, 'A Review of High-Frequency Power Distribution Systems: For Space, Telecommunication, and Computer Applications', *IEEE Trans. Power Electron.*, vol. 29, no. 8, pp. 3852–3863, Aug. 2014.https://doi.org/10.1109/TPEL.2013.2291364
20. K. Acharya, S. K. Mazumder, and I. Basu, 'Reaching Criterion of a Three-Phase Voltage-Source Inverter Operating With Passive and Nonlinear Loads and Its Impact on Global Stability', *IEEE Trans. Ind. Electron.*, vol. 55, no. 4, pp. 1795–1812, Apr. 2008.https://doi.org/10.1109/TIE.2008.917101
21. J. Drobnik, L. Huang, P. Jain, and R. Steigerwald, 'PC platform power distribution system-past application, today's challenge and future direction', in *21st International Telecommunications Energy Conference. INTELEC '99 (Cat. No.99CH37007)*, Copenhagen, Denmark, 1999, p. 36. https://doi.org/10.1109/INTLEC.1999.794011
22. A. Poorfakhraei, M. Narimani, and A. Emadi, 'A Review of Modulation and Control Techniques for Multilevel Inverters in Traction Applications', *IEEE Access*, vol. 9, pp. 24187–24204, 2021.https://doi.org/10.1109/ACCESS.2021.3056612
23. F. Lin, X. Zhang, X. Li, C. Liu, and H. Ma, 'Parameter Design for Symmetrical CLLC-Type DC Transformer Considering Cascaded System Stability and Power Efficiency', *IEEE J. Emerg. Sel. Topics Power Electron.*, pp. 1–1, 2020. https://doi.org/10.1109/JESTPE.2020.3034402
24. F. Blaabjerg, H. Wang, I. Vernica, B. Liu, and P. Davari, 'Reliability of Power Electronic Systems for EV/HEV Applications', *Proc. IEEE*, pp. 1–17, 2020. https://doi.org/10.1109/JPROC.2020.3031041
25. O. M. Hebala, A. A. Aboushady, K. H. Ahmed, and I. Abdelsalam, 'Generic Closed-Loop Controller for Power Regulation in Dual Active Bridge DC–DC Converter With Current Stress Minimization', *IEEE Trans. Ind. Electron.*, vol. 66, no. 6, pp. 4468–4478, Jun. 2019.https://doi.org/10.1109/TIE.2018.2860535
26. J. Huang, Y. Wang, Z. Li, and W. Lei, 'Unified Triple-Phase-Shift Control to Minimize Current Stress and Achieve Full Soft-Switching of Isolated Bidirectional DC–DC Converter', *IEEE Trans. Ind. Electron.*, vol. 63, no. 7, pp. 4169–4179, Jul. 2016.https://doi.org/10.1109/TIE.2016.2543182
27. B. Zhao, Q. Song, and W. Liu, 'Efficiency Characterization and Optimization of Isolated Bidirectional DC–DC Converter Based on Dual-Phase-Shift Control for DC Distribution Application', *IEEE Trans. Power Electron.*, vol. 28, no. 4, pp. 1711–1727, Apr. 2013.https://doi.org/10.1109/TPEL.2012.2210563
28. F. Lin, X. Zhang, X. Li, H. Ma, and C. Cai, 'Design of Symmetrical CLLC Resonant DC Transformer Considering Voltage Transfer Ratio and Cascaded System Stability', *IEEE Trans. Ind. Electron.*, pp. 1–1, 2021. https://doi.org/10.1109/TIE.2020.3048283
29. J. Huang and X. Zhang, 'Step-by-Step Threshold Design Methodology of Switching Frequency Regulation for a Generalized Resonant DC–DC Converter', *IEEE Trans. Ind. Electron.*, vol. 67, no. 12, pp. 10975–10980, Dec. 2020.https://doi.org/10.1109/TIE.2019.2959514
30. B. Zhao, Q. Yu, and W. Sun, 'Extended-Phase-Shift Control of Isolated Bidirectional DC–DC Converter for Power Distribution in Microgrid', *IEEE Trans. Power Electron.*, vol. 27, no. 11, pp. 4667–4680, Nov. 2012.https://doi.org/10.1109/TPEL.2011.2180928
31. A. Tong, L. Hang, G. Li, X. Jiang, and S. Gao, 'Modeling and Analysis of a Dual-Active-Bridge-Isolated Bidirectional DC/DC Converter to Minimize RMS Current With Whole Operating Range', *IEEE Trans. Power Electron.*, vol. 33, no. 6, pp. 5302–5316, Jun. 2018.https://doi.org/10.1109/TPEL.2017.2692276

32. P. Pelletier, J.-M. Guichon, J.-L. Schanen, and D. Frey, 'Optimization of a DC Capacitor Tank', *IEEE Transactions on Industry Applications*, vol. 45, no. 2, pp. 880–886, 2009.https://doi.org/10.1109/TIA.2009.2013579
33. E. I. Amoiralis, M. A. Tsili, D. G. Paparigas, and A. G. Kladas, 'Global Transformer Design Optimization Using Deterministic and Nondeterministic Algorithms', *IEEE Trans. on Ind. Applicat.*, vol. 50, no. 1, pp. 383–394, Jan. 2014.https://doi.org/10.1109/TIA.2013.2288417
34. T. C. Neugebauer and D. J. Perreault, 'Computer-aided optimization of DC/DC converters for automotive applications', *IEEE Trans. Power Electron.*, vol. 18, no. 3, pp. 775–783, May 2003.https://doi.org/10.1109/TPEL.2003.810866
35. S. Balachandran and F. Lee, 'Algorithms for Power Converter Design Optimization', *IEEE Trans. Aerosp. Electron. Syst.*, vol. AES-17, no. 3, pp. 422–432, May 1981. https://doi.org/10.1109/TAES.1981.309070
36. S. Busquets-Monge *et al.*, 'Design of a Boost Power Factor Correction Converter Using Optimization Techniques', *IEEE Trans. Power Electron.*, vol. 19, no. 6, pp. 1388–1396, Nov. 2004.https://doi.org/10.1109/TPEL.2004.836638
37. Y. Tang *et al.*, 'Artificial Intelligence-Aided Minimum Reactive Power Control for the DAB Converter Based on Harmonic Analysis Method', *IEEE Trans. Power Electron.*, pp. 1–1, 2021. https://doi.org/10.1109/TPEL.2021.3059750
38. Keliang Zhou and Danwei Wang, 'Relationship between space-vector modulation and three-phase carrier-based PWM: a comprehensive analysis [three-phase inverters]', *IEEE Trans. Ind. Electron.*, vol. 49, no. 1, pp. 186–196, Feb. 2002.https://doi.org/10.1109/41.982262
39. L. Chen *et al.*, 'Moving Discretized Control Set Model-Predictive Control for Dual-Active Bridge With the Triple-Phase Shift', *IEEE Trans. Power Electron.*, vol. 35, no. 8, pp. 8624–8637, Aug. 2020.https://doi.org/10.1109/TPEL.2019.2962838
40. B. K. Bose, 'Neural Network Applications in Power Electronics and Motor Drives—An Introduction and Perspective', *IEEE Trans. Ind. Electron.*, vol. 54, no. 1, pp. 14–33, Feb. 2007.https://doi.org/10.1109/TIE.2006.888683
41. J. Y. Goulermas, P. Liatsis, Xiao-Jun Zeng, and P. Cook, 'Density-Driven Generalized Regression Neural Networks (DD-GRNN) for Function Approximation', *IEEE Trans. Neural Netw.*, vol. 18, no. 6, pp. 1683–1696, Nov. 2007. https://doi.org/10.1109/TNN.2007.902730
42. O. I. Abiodun *et al.*, 'Comprehensive Review of Artificial Neural Network Applications to Pattern Recognition', *IEEE Access*, vol. 7, pp. 158820–158846, 2019.https://doi.org/10.1109/ACCESS.2019.2945545
43. Z. Zou, Z. Shi, Y. Guo, and J. Ye, 'Object Detection in 20 Years: A Survey', arXiv:1905.05055 *[cs]*, May 2019, Accessed: Jun. 19, 2021. [Online]. Available: http://arxiv.org/abs/1905.05055
44. S. Wen, Y. Wang, Y. Tang, Y. Xu, P. Li, and T. Zhao, 'Real-Time Identification of Power Fluctuations Based on LSTM Recurrent Neural Network: A Case Study on Singapore Power System', *IEEE Trans. Ind. Inf.*, vol. 15, no. 9, pp. 5266–5275, Sep. 2019.https://doi.org/10.1109/TII.2019.2910416
45. S. E. De Leon-Aldaco, H. Calleja, and J. Aguayo Alquicira, 'Metaheuristic Optimization Methods Applied to Power Converters: A Review', *IEEE Trans. Power Electron.*, vol. 30, no. 12, pp. 6791–6803, Dec. 2015. https://doi.org/10.1109/TPEL.2015.2397311
46. Yaow-Ming Chen, 'Passive filter design using genetic algorithms', *IEEE Transactions on Industrial Electronics*, vol. 50, no. 1, pp. 202–207, Feb. 2003.https://doi.org/10.1109/TIE.2002.807664
47. X. Li, K. Mao, F. Lin, and X. Zhang, 'Particle swarm optimization with state-based adaptive velocity limit strategy', *Neurocomputing*, vol. 447, pp. 64–79, Aug. 2021.https://doi.org/10.1016/j.neucom.2021.03.077
48. T. Liao, K. Socha, M. A. Montes de Oca, T. Stutzle, and M. Dorigo, 'Ant Colony Optimization for Mixed-Variable Optimization Problems', *IEEE Trans. Evol. Computat.*, vol. 18, no. 4, pp. 503–518, Aug. 2014.https://doi.org/10.1109/TEVC.2013.2281531
49. C. A. Coello Coello, 'Evolutionary multi-objective optimization: a historical view of the field', *IEEE Comput. Intell. Mag.*, vol. 1, no. 1, pp. 28–36, Feb. 2006. https://doi.org/10.1109/MCI.2006.1597059

50. L. A. Zadeh, 'Fuzzy logic = computing with words', *IEEE Trans. Fuzzy Syst.*, vol. 4, no. 2, pp. 103–111, May 1996.https://doi.org/10.1109/91.493904
51. S. Guillaume, 'Designing fuzzy inference systems from data: An interpretability-oriented review', *IEEE Trans. Fuzzy Syst.*, vol. 9, no. 3, pp. 426–443, Jun. 2001.https://doi.org/10.1109/91.928739
52. Hao Ying, Yongsheng Ding, Shaokuan Li, and Shihuang Shao, 'Comparison of necessary conditions for typical Takagi-Sugeno and Mamdani fuzzy systems as universal approximators', *IEEE Trans. Syst., Man, Cybern. A*, vol. 29, no. 5, pp. 508–514, Sep. 1999. https://doi.org/10.1109/3468.784177
53. T. Dragicevic, P. Wheeler, and F. Blaabjerg, 'Artificial Intelligence Aided Automated Design for Reliability of Power Electronic Systems', *IEEE Trans. Power Electron.*, vol. 34, no. 8, pp. 7161–7171, Aug. 2019.https://doi.org/10.1109/TPEL.2018.2883947
54. X. Li, X. Zhang, F. Lin, and F. Blaabjerg, 'Artificial-Intelligence-Based Design (AI-D) for Circuit Parameters of Power Converters', *IEEE Trans. Ind. Electron.*, pp. 1–1, 2021. https://doi.org/10.1109/TIE.2021.3088377
55. J. Song, F. Dong, J. Zhao, H. Wang, Z. He, and L. Wang, 'An Efficient Multiobjective Design Optimization Method for a PMSLM Based on an Extreme Learning Machine', *IEEE Trans. Ind. Electron.*, vol. 66, no. 2, pp. 1001–1011, Feb. 2019.https://doi.org/10.1109/TIE.2018.2835413
56. F. Filho, H. Z. Maia, T. H. A. Mateus, B. Ozpineci, L. M. Tolbert, and J. O. P. Pinto, 'Adaptive Selective Harmonic Minimization Based on ANNs for Cascade Multilevel Inverters With Varying DC Sources', *IEEE Trans. Ind. Electron.*, vol. 60, no. 5, pp. 1955–1962, May 2013.https://doi.org/10.1109/TIE.2012.2224072
57. F.-J. Lin, K.-C. Lu, and B.-H. Yang, 'Recurrent Fuzzy Cerebellar Model Articulation Neural Network Based Power Control of a Single-Stage Three-Phase Grid-Connected Photovoltaic System During Grid Faults', *IEEE Trans. Ind. Electron.*, vol. 64, no. 2, pp. 1258–1268, Feb. 2017.https://doi.org/10.1109/TIE.2016.2618882
58. M. P. Kazmierkowski, L. E. B. Da Silva, B. K. Bose, and J. O. P. Pinto, 'A neural-network-based space-vector PWM controller for voltage-fed inverter induction motor drive', *IEEE Trans. on Ind. Applicat.*, vol. 36, no. 6, pp. 1628–1636, Dec. 2000.https://doi.org/10.1109/28.887215
59. S. Kiranyaz, A. Gastli, L. Ben-Brahim, N. Al-Emadi, and M. Gabbouj, 'Real-Time Fault Detection and Identification for MMC Using 1-D Convolutional Neural Networks', *IEEE Trans. Ind. Electron.*, vol. 66, no. 11, pp. 8760–8771, Nov. 2019.https://doi.org/10.1109/TIE.2018.2833045
60. Shoou-Jinn Chang, Hao-Sheng Hou, and Yan-Kuin Su, 'Automated passive filter synthesis using a novel tree representation and genetic programming', *IEEE Trans. Evol. Computat.*, vol. 10, no. 1, pp. 93–100, Feb. 2006.https://doi.org/10.1109/TEVC.2005.861415
61. G. Adinolfi, G. Graditi, P. Siano, and A. Piccolo, 'Multiobjective Optimal Design of Photovoltaic Synchronous Boost Converters Assessing Efficiency, Reliability, and Cost Savings', *IEEE Trans. Ind. Inf.*, vol. 11, no. 5, pp. 1038–1048, Oct. 2015.https://doi.org/10.1109/TII.2015.2462805
62. Y. Liu, M. Huang, H. Wang, X. Zha, J. Gong, and J. Sun, 'Reliability-Oriented Optimization of the LC Filter in a Buck DC-DC Converter', *IEEE Transactions on Power Electronics*, vol. 32, no. 8, pp. 6323–6337, Aug. 2017.https://doi.org/10.1109/TPEL.2016.2619690
63. S. Coupland and R. John, 'Geometric Type-1 and Type-2 Fuzzy Logic Systems', *IEEE Trans. Fuzzy Syst.*, vol. 15, no. 1, pp. 3–15, Feb. 2007.https://doi.org/10.1109/TFUZZ.2006.889764
64. Y. H. Kim, S. C. Ahn, and W. H. Kwon, 'Computational complexity of general fuzzy logic control and its simplification for a loop controller', *Fuzzy Sets and Systems*, vol. 111, no. 2, pp. 215–224, Apr. 2000.https://doi.org/10.1016/S0165-0114(97)00409-0
65. G. Durgasukumar and M. K. Pathak, 'Comparison of adaptive Neuro-Fuzzy-based space-vector modulation for two-level inverter', *International Journal of Electrical Power & Energy Systems*, vol. 38, no. 1, pp. 9–19, Jun. 2012.https://doi.org/10.1016/j.ijepes.2011.10.017
66. X. Li, X. Zhang, F. Lin, C. Sun, and K. Mao, "Artificial-Intelligence-Based Triple Phase Shift Modulation for Dual Active Bridge Converter with Minimized Current Stress," *IEEE J. Emerg. Sel. Top. Power Electron.*, pp. 1–1, 2021. https://doi.org/10.1109/JESTPE.2021.3105522

67. H. Abu-Rub, A. Iqbal, Sk. Moin Ahmed, F. Z. Peng, Y. Li, and G. Baoming, 'Quasi-Z-Source Inverter-Based Photovoltaic Generation System With Maximum Power Tracking Control Using ANFIS', *IEEE Trans. Sustain. Energy*, vol. 4, no. 1, pp. 11–20, Jan. 2013. https://doi.org/10.1109/TSTE.2012.2196059
68. J. Saroha, M. Singh, and D. K. Jain, 'ANFIS-Based Add-On Controller for Unbalance Voltage Compensation in a Low-Voltage Microgrid', *IEEE Trans. Ind. Inf.*, vol. 14, no. 12, pp. 5338–5345, Dec. 2018.https://doi.org/10.1109/TII.2018.2803748

Chapter 2
The Proposed Automated Optimal Design for Power Switch: A Thermo-mechanical-Coordinated and Multi-objective-Oriented Optimization Methodology

Check for updates

2.1 Introduction

With the rising popularity of electric vehicles (EVs), the power control unit (PCU) of EVs has gained widespread attention [1]. As shown in Fig. 2.1, the PCU regulates the power flow between battery and motor, which is commonly regarded as the heart of EV. Aside from the battery, the power module used in the PCU is the most costly component of the EV, accounting for 7–15% of the total cost [2]. To meet the harsh mission profile and match the lifetime expectancy of the EV, the power module must meet the requirements of low heat resistance and mechanical stress toward high reliability and durability (in general, 1 million km or 10 years lifetime) [3, 4]. The emerging double-sided cooling (DSC) power module provides better thermal dissipation performance and reduced packaging parasitics as compared to the conventional single-sided cooling (SSC) power module. As shown in Fig. 2.1, the DSC power module has recently been employed more frequently in EVs to enhance the PCU's long-term reliability, power density, and energy efficiency [5, 6]. The emerging DSC power module has not, however, been widely commercialized in opposition to the mature SSC power module during the last decades because of a lack of design methodology, ambiguous thermo-mechanical stress, an uncertain lifespan principle, etc. These problems block the DSC power module's large-scale supply.

The DSC power module can be subdivided into low-temperature co-fired ceramics (LTCC) [7], pressure contact [8–10], and direct soldering [11, 12] by the top side interconnection ways of dies. The DSC power module for EV applications typically has a rating between 600 and 1200 V. Considering the cost issue, the soldered DSC power module is mostly used.

Recent research has made an effort to enhance the DSC power module from the perspectives of reduced thermal resistance, decreased parasitic inductance, and improved mechanical strength.

X. Zhang et al., *Automated Design of Electrical Converters with Advanced AI Algorithms*,
https://doi.org/10.1007/978-981-99-0459-4_2

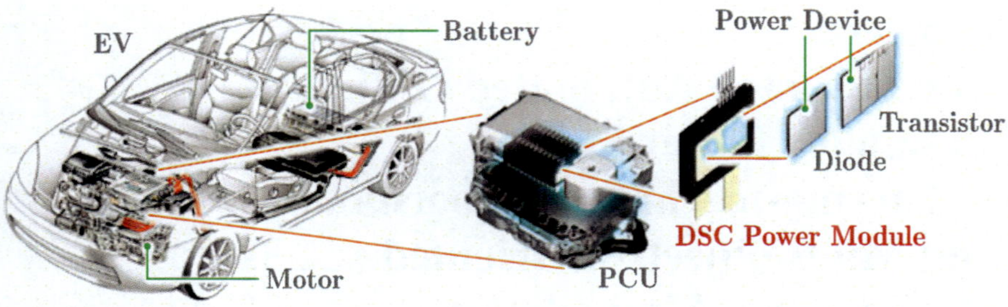

Fig. 2.1 Implementation of DSC power module in PCU of EV

- In terms of parasitic inductance reduction: Some novel packaging structures are developed. The P-cell and N-cell concepts are presented to minimize commutation parasitics and shorten the power loop [13, 14]. By employing alternating displacement of the transistor and diode, the split P- and N-cell layout reduces the parasitic inductance of the DSC power module by more than 50% when compared to the conventional layout. Moreover, the folded overlay power loop is proposed to decrease the parasitic inductance of the DSC power module and minimize the magnetic fields [14, 15]. The mutual inductances cancel each other in these improved DSC power modules, and the commutation area is limited to the width of dies in the power loop. Additionally, it is proposed to constrain the commutation area within the thickness of dies using the chip-on-chip stack and flip-chip packaging concepts [16–19]. As a consequence, it is possible to adjust the parasitic inductance well within 2 nH. The DSC power module can be subdivided into low-temperature co-fired ceramics (LTCC) [7], pressure contact [8–10], and direct soldering [11, 12] by the top side interconnection ways of dies. The DSC power module for EV applications typically has a rating between 600 and 1200 V. Considering the cost issue, the soldered DSC power module is mostly used.
- In terms of thermal resistance decrease: There are some novel packaging materials used. To replace the conventional SnAgCu (SAC) alloy solder and reduce thermal resistance during die-attachment, high thermal conductivity nano-silver sintering has been proposed [20, 21]. The synthetic diamond is anticipated to replace the ceramic for the directly bonded coppers (DBC) substrate due to its high thermal conductivity and matching coefficient thermal expansion (CTE) near the semiconductor material [22]. The use of aluminum-based bumps in the flip-chip mounting technique is proposed to improve the thermal performance of the DSC power module. Compared with the SSC power module, the DSC power module's thermal capacity improves by 1.7%, and its transient thermal impedance decreases by 15% [23]. The spacer is designed to strike a compromise between lowering heat resistance and enhancing electrical insulation [24]. Some novel structures are developed for the spacers, including bumps, solid posts, and the embedded substrate. Instead of copper (Cu), the metal molybdenum (Mo) is suggested for the spacer. The assembled heat-sink is incorporated into the DSC power module to improve heat dissipation [25]. Some optimized cooling approaches, including

advanced spraying cooling and cutting-edge micro-channel cooling, are applied to lower the DSC power module's junction-ambient thermal resistance [26, 27].

- A finite element analysis technique is employed to examine the warpage behavior of the DSC power module with alternative DSC packaging structures [28]. The mismatched CTEs of the stacked layers may cause a bi-metal-like bow effect [29] due to the heterogeneous integration of the DSC power module, which raises junction-heatsink thermal resistance and results in the chip overheating. To address the possibility of flexing or bowing, the DSC module's symmetrical stack is proposed [30]. The bumps constructed with molybdenum and copper are proposed to reduce the maximum residual stress and the creep strain accumulation at the corners of the solder layer [31]. A quick reliability test methodology is described in order to emulate the mechanical stresses caused by thermal expansion during thermal cycling of the DSC power module and mitigate the shear stress caused by thermal effects [32].

Although conventional designs are beneficial in improving the performance of the DSC power module by using better structures and materials, it is heavily reliant on trial-and-error [33, 34]. It is time-consuming and expensive, undermining its market competitiveness. Additionally, the conventional design is prone to parasitic metrics; nevertheless, the thermo-mechanical metrics are not given enough attention throughout the design phase. Furthermore, the metrics are not mathematically modelled in the current design of the DSC power module. As a consequence, it is also impossible to design automatically [35, 36].

To summarize, the conventional DSC power module design lacks a conceptual model and design methodology to coordinate thermo-mechanical metrics and fulfill automated design requirements. Furthermore, the conventional design simply takes into account a single measure and overlooks the tradeoff between multi-physics metrics. As a result, it is necessary to simulate how material properties and structural sizes affect the thermo-mechanical performance of the DSC power module. Furthermore, how to balance the thermo-mechanical tradeoff and optimize the DSC power module should be emphasized.

This chapter intends to fill these research gaps by proposing a thermo-mechanically coordinated multi-objective optimization method that employs the evolutionary algorithm to autonomously design the DSC power module. The remainder of this chapter is arranged as follows. In Sect. 2.2, the present practices and the under-optimization problem of the DSC power module are proposed. In Sect. 2.3, mathematical models of the thermal resistance and mechanical stress in the DSC power module are proposed. Meanwhile, the thermo-mechanical-driven multi-objective optimization model is proposed. In Sect. 2.4, the Pareto solutions are achieved for optimal structure sizes of the DSC power module by using an evolutionary algorithm. Meanwhile, the influences of material properties on the optimal solutions are illustrated. Prototype and experiment results are demonstrated in Sect. 2.5. Finally, some conclusions are drawn in Sect. 2.6.

2.2 Present Status and Under-Optimization Problem of DSC Power Module

2.2.1 State-of-the-Art of DSC Power Module

As demonstrated in Fig. 2.2, several renowned enterprises and organizations exploited the feasibility of the DSC power module [7–32]. These cutting-edge DSC power modules vary from 50 to 600 A and alter from single-switch to six-in-one assembly. Apart from the prototypes, the pioneer DSC power modules from Denso and Infineon will be commercialized for industrial-scale deployment. The DSC power module is becoming more popular for EV applications in recent years.

Taking the latest commercial DSC power module FF400R07A01E3_S6 for example, its configuration is demonstrated in Fig. 2.3a and b. It is a 400 A/700 V half-bridge DSC power module from Infineon. Cross-section structure of the power module is shown in Fig. 2.3c [37, 38]. It consists of two Si IGBT transistors, two Si FRD diodes, two DBCs, and several spacers soldered on the top side of the dies. The dies are protected by high-temperature and high-performance molding resin.

Based on Fig. 2.3b, the general cross-section cell of the DSC power module in the x–z plane is illustrated in Fig. 2.4. It is a complex multi-layer element, including two DBCs, a spacer, a chip, and three solder attachments.

DBC is a copper-ceramic-copper sandwich structure for thermal conduction and isolation galvanic. The spacer is employed to provide sufficient electric insulation and margin height for the transistor's gate pad bonding wire. Solder attachments are utilized to connect the heterogeneous layers. Without a doubt, once the materials are decided, the performances of the DSC power module are heavily dependent on the sizes of the heterogeneous layers.

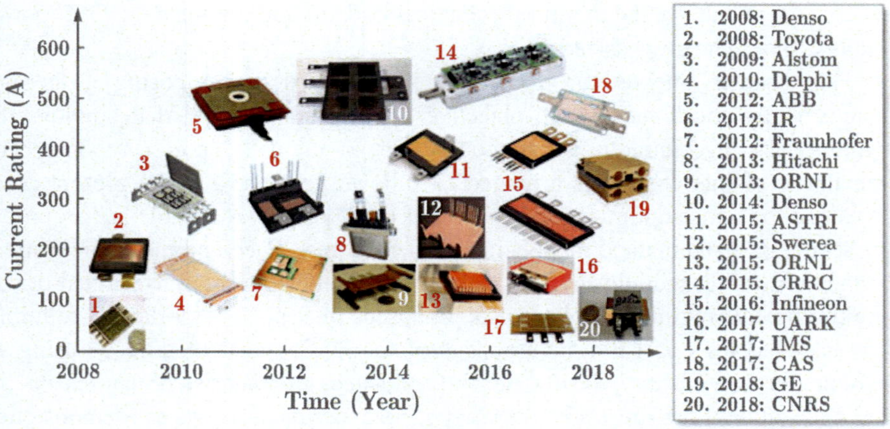

Fig. 2.2 State-of-the-art of DSC power modules

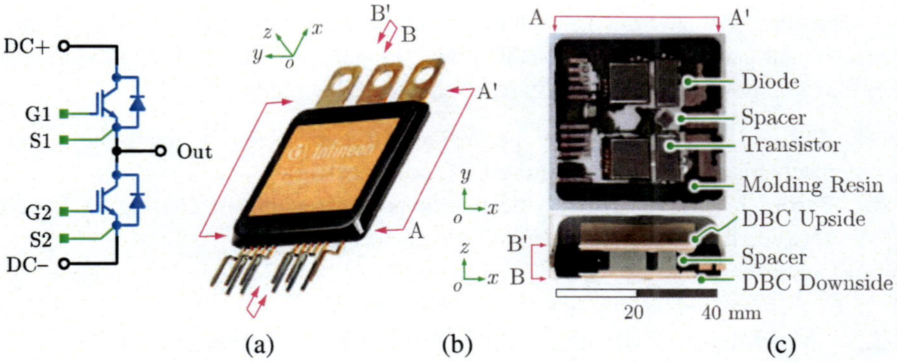

Fig. 2.3 Commercial DSC power module. **a** Configuration, **b** overview, and **c** cross-section

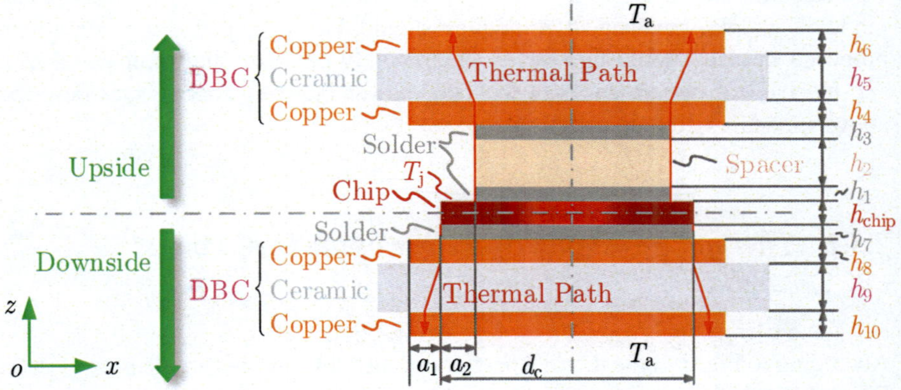

Fig. 2.4 Schematic cross-section of DSC power module

2.2.2 Under-Optimization Problem of DSC Power Module

As illustrated in Figs. 2.3 and 2.4, the heights of DBCs on the upside and downside for the commercial DSC power module are the same, taking the ceramic layers of DBCs for instance, $h_5 = h_9$. However, since the spacer layer is only on the top, the DSC power module's vertical structure is asymmetric. As a result, for an ideal DSC power module, the heights of layers on the upside and downside may not be consistent (for example, $h_5 \neq h_9$). Clearly, the present DSC power module is under-optimal, and it should be further modeled and improved to meet certain targets.

Based on Fig. 2.4, the objectives, variables, and limitations for optimizing the DSC power module are given as follows. Considering the high-reliability requirement of EV implementation, the lifetime-dominated thermal resistance and mechanical stress are selected as objectives. Heights h_1 to h_{10}, as well as widths a_1 and a_2, are considered as the variables to be optimized. Some manufacturing limitations are considered as constraints of the variables.

The structure sizes and material properties have a profound effect on the thermo-mechanical performance of the DSC power module. Aiming at the optimal design of the DSC power module, two barriers should be overcome.

(a) Barrier 1: How material properties and structure sizes dominate the thermo-mechanical performance of the DSC power module.
(b) Barrier 2: How to optimally design the power module considering the tradeoff between thermal and mechanical specifications.

2.3 Modeling of the Proposed Thermo-mechanical Multi-objective Co-design of DSC Power Module

To overcome the obstacles in Sect. 2.2, mathematical models of the DSC power module's thermo-mechanical principles should be developed. Furthermore, the proposed models should be used to coordinately optimize and steer the thermal and mechanical performances of the DSC power module. These issues described above are addressed in this section.

2.3.1 *Thermal Resistance Modeling of the DSC Power Module*

According to Fig. 2.4, the thermal network of the DSC power module is proposed in Fig. 2.5. P_L is the power loss of the chip. T_j and T_a are the junction temperature and ambient temperature. Due to the asymmetric structure of the DSC power module in the vertical direction, thermal resistances on the upside and downside are unequal.

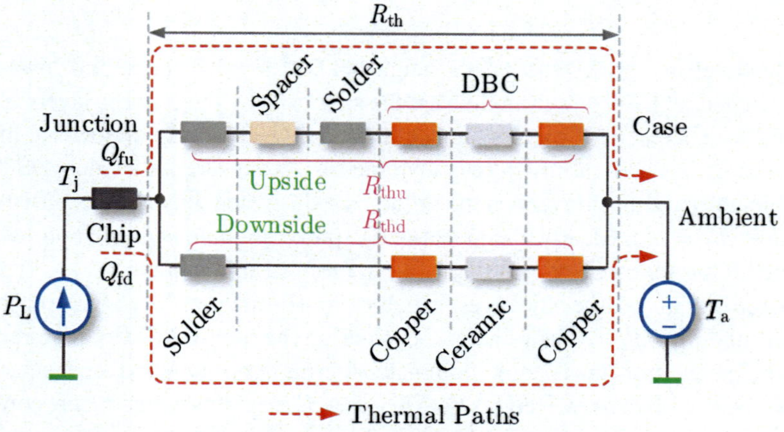

Fig. 2.5 Thermal network model of the DSC power module

Based on Fig. 2.5, the junction-case thermal resistance can be proposed as

$$R_{\text{thjc}} = \frac{R_{\text{thjcu}} R_{\text{thjcd}}}{R_{\text{thjcu}} + R_{\text{thjcd}}} \tag{2.1}$$

where $R_{\text{th}u}$ and $R_{\text{th}d}$ are respective thermal resistances of thermal paths on the upside and downside, which can be modelled as

$$\begin{cases} R_{\text{thu}} = \sum_{i=1}^{6} R_{\text{th}i} = \sum_{i=1}^{6} \frac{h_i}{\lambda_i A_i} \\ R_{\text{thd}} = \sum_{i=7}^{10} R_{\text{th}i} = \sum_{i=7}^{10} \frac{h_i}{\lambda_i A_i} \end{cases} \tag{2.2}$$

where $R_{\text{th}i}$, λ_i, and A_i are thermal resistance, thermal conductivity, and equivalent thermal conduction area of the ith layer. According to Fig. 2.4, areas of the double-sided die-attach solders, A_1 and A_7, can be calculated by $A_1 = (d_c - 2a_2)^2$ and $A_7 = {d_c}^2$. d_c is the chip width. Considering the thermal spreading effect, the equivalent areas of other layers can be defined as

$$\begin{cases} A_{i+1} = \left(\sqrt{A_i} + \sum_{n=1}^{i} \frac{\lambda_n}{\lambda_{n+1}} h_{n+1}\right)^2 & 1 \le i \le 5 \\ A_{i+1} = \left(\sqrt{A_i} + \sum_{n=7}^{i} \frac{\lambda_n}{\lambda_{n+1}} h_{n+1}\right)^2 & 7 \le i \le 9 \end{cases} \tag{2.3}$$

The thermal flux through the parallel thermal paths of the DSC power module can be expressed as

$$\begin{cases} Q_{fu} = \frac{P_{fu}}{A_1} = \frac{R_{thu} \Delta T_j}{A_1} \\ Q_{fd} = \frac{P_{fd}}{A_7} = \frac{R_{thd} \Delta T_j}{A_7} \end{cases} \tag{2.4}$$

where Q_{fu} and Q_{fd} are the thermal fluxes through upside and downside, respectively. P_{fu} and P_{fd} are the dissipation losses conducted by the upside and downside, $P_{fu} + P_{fd} = P_L$. The thermal flux dispatching on the parallel thermal paths can be written as

$$\begin{cases} \frac{Q_{fu}}{Q_{fd}} = \frac{R_{thu} A_7}{R_{thd} A_1} \approx \frac{R_{thu}}{R_{thd}} \\ \frac{Q_{fu}}{Q_{fu} + Q_{fd}} = \frac{R_{thu} A_7}{R_{thu} A_7 + R_{thd} A_1} \approx \frac{R_{th}}{R_{thd}} \end{cases} \tag{2.5}$$

It means the thermal path has a less thermal resistance and proportionally conducts more thermal flux.

Typical material properties of the DSC power module are summarized in Table 2.1. Widely used solders for attachment are SAC305 (Sn96.5/Ag3/Cu0.5) alloy,

Table 2.1 Properties of materials for power module

Layer	Materials	λ [W/(m·K)]	β (ppm/K)	E (GPa)
Solder	SAC305	63.2	21.6	50
	Sn63Pb37	52.8	23.3	40
	Ag Paste	429	19.5	85
Ceramic	Al_2O_3	24	6.5	400
	AlN	180	4.5	310
	Si_3N_4	90	3.3	250
Metal	Mo	139	4.8	329
	Cu	380	17	141
	Ag	406	18	69

Sn63Pb37 alloy, and silver (Ag) paste. Commonly used ceramics for the DBC are Al_2O_3, AlN, and Si_3N_4. Usually used metals for the spacer are Mo, Cu, and Ag.

Typical material properties and structure sizes of the under-optimization DSC power module are listed in Table 2.2. The chip width is $d_c = 4.25$ mm. Except for the isolation rings and the gate pad, the active area of the topside die usable for soldering is 3.5 mm × 3.5 mm. The actual width of the ith layer is d_i. The edge margin widths are $a_1 = 0.875$ mm and $a_2 = 0.375$ mm.

The heat distribution of the DSC power module is represented in Fig. 2.6 using the finite element analysis (FEA) according to the parameters in Table 2.2. To simulate the water cooling environment, a thermal exchange coefficient of 6 kW/(m^2·K) is set on the topside and downside surfaces of the DSC power module. The DSC power module's junction-case thermal resistance is determined to be 0.202 K/W. Furthermore, 15% of the thermal flux is shown to flow through the top DBC. The topside die has a smaller heat dissipation area than the downside die. Furthermore, the spacer and extra solder layer are embedded on the topside of the thermal route.

Table 2.2 Typical sizes of DSC power module

Layer		Materials	Height (h_i) (mm)	Width (d_i) (mm)
Top side	1	Solder (SAC305)	0.1	3.5
	2	Spacer (Mo)	2	3.5
	3	Solder (SAC305)	0.1	3.5
	4	Copper (Cu)	0.2	6
	5	Ceramic (AlN)	0.3	8
	6	Copper (Cu)	0.2	6
Down side	7	Solder (SAC305)	0.1	4.25
	8	Copper (Cu)	0.2	6
	9	Ceramic (AlN)	0.3	8
	10	Copper (Cu)	0.2	6

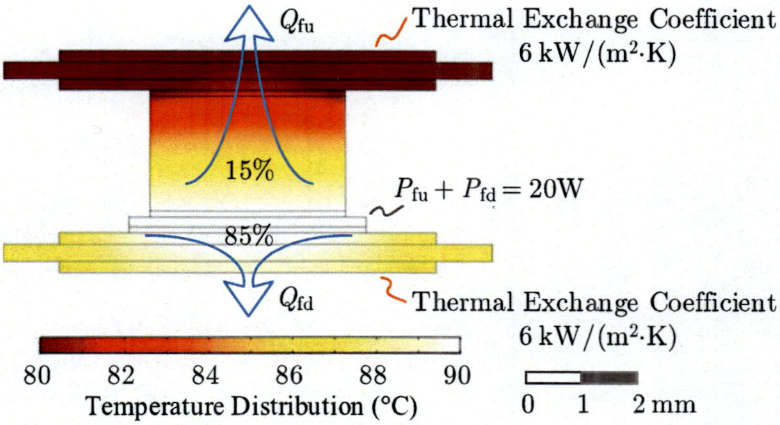

Fig. 2.6 Thermal distribution of DSC power module by using FEA

As a result, the topside has more thermal resistance than the downside. The thermal resistance of the DSC power module is 15% lower than that of the conventional SSC power module.

According to the proposed models in (2.1)–(2.3) and the packaging data in Tables 2.1 and 2.2, the breakdown thermal resistance of the DSC power module in Fig. 2.4 is illustrated in Fig. 2.7. The FEA is employed to understand the thermal resistance principle of the DSC power module. Compared with the estimated thermal resistance of 0.202 K/W by using FEA, the theoretical result according to (2.1)–(2.3) is 0.208 K/W having a relative error of 2.8%. Obviously, the proposed models are accurate enough to characterize the thermal resistance of the DSC power module.

Furthermore, as shown in Fig. 2.7, the spacer and solder layers have a significant influence on the thermal resistance of the DSC power module. The thermal resistances affected by the solder and spacer layers are shown in Fig. 2.8 using various heights

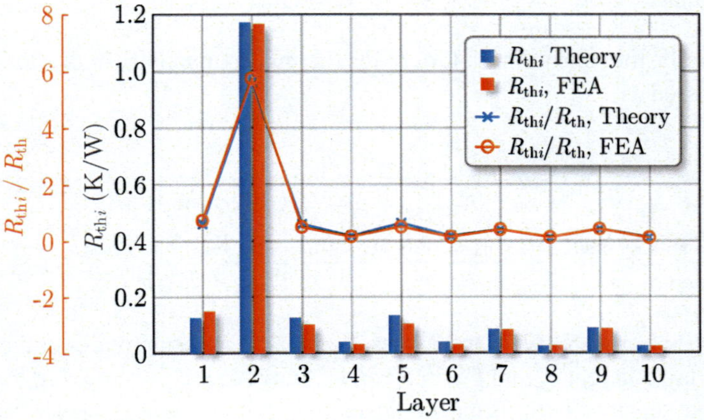

Fig. 2.7 Breakdown thermal resistance of DSC power module

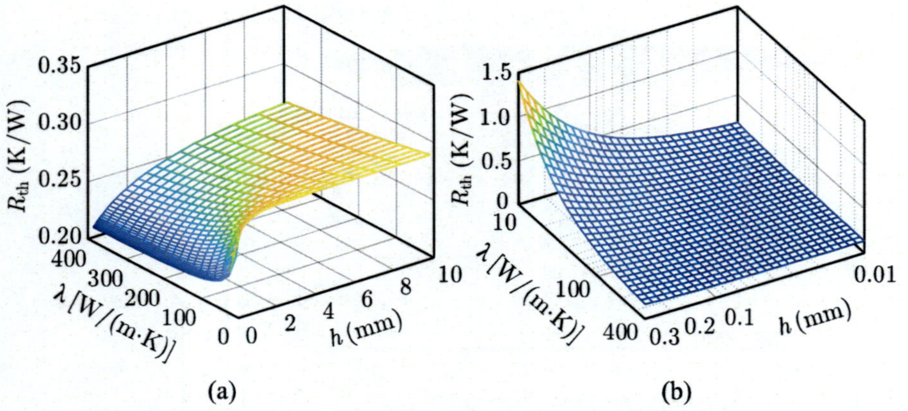

Fig. 2.8 Thermal resistance of DSC power module influenced by **a** spacer and **b** solder

and materials. The thin size and high thermal conductivity material should be applied to the spacer and solder layers in order to reduce the junction-case thermal resistance.

Furthermore, due to the low thermal conductivity of metal Mo, the thermal resistance of the spacer is a bottleneck of the DSC power module. By using high thermal conductivity metals like Cu, the thermal resistance of the spacer can be reduced by 63%. However, the coefficient of thermal expansion (CTE) β of Cu is larger than Mo, which remarkably increases the mechanical stress of the DSC power module to be illustrated in the next subsection.

According to (2.1)–(2.3), the sensitivities of thermal resistance R_{th} to the height of the ith layer h_i can be derived as

$$\begin{cases} \dfrac{\partial R_{\mathrm{th}}}{\partial h_i} = \dfrac{\partial R_{\mathrm{thu}}}{\partial h_i}\dfrac{R_{\mathrm{thd}}^2}{(R_{\mathrm{thu}}+R_{\mathrm{thd}})^2} = \dfrac{1}{\lambda_i A_i}\dfrac{R_{\mathrm{thd}}^2}{(R_{\mathrm{thu}}+R_{\mathrm{thd}})^2}, & 1 \le i \le 6 \\ \dfrac{\partial R_{\mathrm{th}}}{\partial h_i} = \dfrac{\partial R_{\mathrm{thd}}}{\partial h_i}\dfrac{R_{\mathrm{thu}}^2}{(R_{\mathrm{thu}}+R_{\mathrm{thd}})^2} = \dfrac{1}{\lambda_i A_i}\dfrac{R_{\mathrm{thu}}^2}{(R_{\mathrm{thu}}+R_{\mathrm{thd}})^2}, & 7 \le i \le 10 \end{cases} \tag{2.6}$$

Similarly, the sensitivities of R_{th} to the thermal conductivity of the ith layer λ_i can be expressed as

$$\begin{cases} \dfrac{\partial R_{\mathrm{th}}}{\partial \lambda_i} = \dfrac{\partial R_{\mathrm{thu}}}{\partial \lambda_i}\dfrac{R_{\mathrm{thd}}^2}{(R_{\mathrm{thu}}+R_{\mathrm{thd}})^2} = -\dfrac{h_i}{\lambda_i^2 A_i}\dfrac{R_{\mathrm{thd}}^2}{(R_{\mathrm{thu}}+R_{\mathrm{thd}})^2}, & 1 \le i \le 6 \\ \dfrac{\partial R_{\mathrm{th}}}{\partial \lambda_i} = \dfrac{\partial R_{\mathrm{thd}}}{\partial \lambda_i}\dfrac{R_{\mathrm{thu}}^2}{(R_{\mathrm{thu}}+R_{\mathrm{thd}})^2} = -\dfrac{h_i}{\lambda_i^2 A_i}\dfrac{R_{\mathrm{thu}}^2}{(R_{\mathrm{thu}}+R_{\mathrm{thd}})^2}, & 7 \le i \le 10 \end{cases} \tag{2.7}$$

According to (2.6)–(2.7) and Tables 2.1 and 2.2, the sensitivities of thermal resistance to layer height h_i and thermal conductivity λ_i, are illustrated in Fig. 2.9. The layer most sensitive to R_{th} is the die-attachment on the downside (layer 7 in Fig. 2.4). For layer 7, the height increment $\Delta h_7 = 0.1$ mm leads to thermal resistance growth

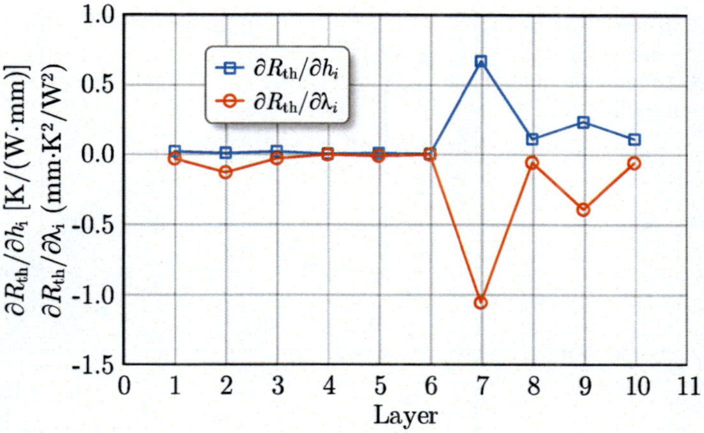

Fig. 2.9 Sensitivities of thermal resistance of DSC power module

$\Delta R_{th} = \frac{\partial R_{th}}{\partial h_7} \Delta h_7 = 0.07$ K/W. However, the increased thermal conductivity $\Delta\lambda_7 =$ 100 W/(m·K) reduces the thermal resistance by $\Delta R_{th} = \frac{\partial R_{th}}{\partial \lambda_7} \Delta\lambda_7 = -0.11$ K/W. In general, the sensitivities of structure height and thermal conductivity are in opposing directions. To minimize R_{th} of DSC power module, low structure heights and high thermal conductivity materials are preferred.

The electro-thermal coupling of the DSC power module contributes to the power cycle lifetime. The power cycling lifetime of the DSC power module is dictated by the junction temperature cycle, which is directly determined by R_{th}, according to the Coffin-Manson theory [39, 40]. In general, the DSC power module's thin layer sizes and high thermal conductivity materials can decrease R_{th} and dramatically improve power cycle lifetime.

According to the Coffin-Manson theory, the power cycling lifetime N_{p} of the DSC power module depends on junction temperature cycle ΔT_{j}, which is proposed as

$$N_{\text{p}} = c_1 (\Delta T_{\text{j}})^{c_2} e^{E_{\text{a}}/(kT_{\text{jm}})} \tag{2.8}$$

where c_1 and c_2 are initial lifetime and accelerated aging factors. T_{jm} is the average junction temperature. E_{a} is the activation energy. k is the Boltzmann constant.

Power cycling lifetime involves the electro-thermal coupling of the DSC power module. Junction temperature cycle ΔT_{j} is determined by R_{th} and ΔP_{L}, which can be expressed as

$$\Delta T_{\text{j}} = R_{\text{th}} \Delta P_{\text{L}} \tag{2.9}$$

where $\Delta P = P_{\text{Lmax}} - P_{\text{Lmin}}$, P_{Lmax}, and P_{Lmin} are peak and valley losses in a power cycle. As a result, the (2.8) can be rewritten as

$$N_{\text{p}} = c_1 (R_{\text{th}} \Delta P_{\text{L}})^{c_2} e^{E_{\text{a}}/(kT_{\text{jm}})} \tag{2.10}$$

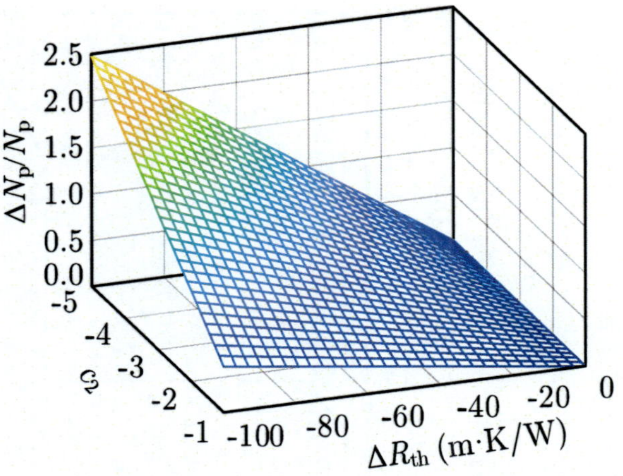

Fig. 2.10 Lifetime increasing by decreasing thermal resistance of the DSC power module

Obviously, lifetime N_p is directly determined by R_th. Small and thin layers of the DSC power module can effectively reduce the R_th and improve the N_p. According to (2.10), the sensitivity of N_p to the thermal resistance can be expressed as

$$\frac{\partial N_\text{p}}{\partial R_\text{th}} = c_1 c_2 R_\text{th}^{c_2-1}(\Delta P_\text{L})^{c_2} e^{E_\text{a}/(kT_\text{jm})} \tag{2.11}$$

By decreasing thermal resistance, the lifetime of the DSC power module can be improved by

$$\frac{\Delta N_\text{p}}{N_\text{p}} = \frac{1}{N_\text{p}}\frac{\partial N_\text{p}}{\partial R_\text{th}}\Delta R_\text{th} = c_2\frac{\Delta R_\text{th}}{R_\text{th}} \tag{2.12}$$

The effect of the thermal resistance on the lifetime of the DSC power module is demonstrated in Fig. 2.10. By using optimal structure sizes and material properties, the decreased thermal resistance can linearly promote the lifetime of the DSC power module.

2.3.2 *Mechanical Stress Modeling of the DSC Power Module*

Power module failures caused by the mechanical stress of the thermal cycling depend on volume-averaged inelastic energy density (VAIED) W. According to the Darveaux theory [41], the lifetime model depended on the mechanical stress can be proposed as

$$\begin{cases} N_\text{t0} = \lambda_1 W^{\lambda_2} \\ \text{d}\theta/\text{d}N_\text{t} = \lambda_3 W^{-\lambda_4} \end{cases} \tag{2.13}$$

where N_{t0} is the initiation number of thermal cycles when a crack appears. θ is the length of the crack. λ_1 to λ_4 are constants. The thermal cycling lifetime N_t can be rewritten as

$$N_t = N_{t0} + \frac{\theta_c}{d\theta/dN_t} = \lambda_1 W^{\lambda_2} + \frac{\theta_c}{\lambda_3} W^{\lambda_4} \quad (2.14)$$

where θ_c is the length of the characteristic crack. Generally, a small W is required to improve the lifetime N_t.

To extend lifetime N_t, how material properties and structure sizes of the power module affect the N_t should be further addressed. The VAIED W can be expressed as

$$W = \frac{\sum_{i=1}^{10} W_i V_i}{\sum_{i=1}^{10} V_i} \quad (2.15)$$

where W_i and V_i represent the inelastic energy density and volume of the ith layer, respectively. V_i can be computed by $V_i = S_i h_i$, where $S_i = d_i{}^2$ is the actual area of the ith layer. Additionally, the W_i can be modelled as

$$W_i = \frac{F_i \Delta h_i}{V_i} \quad (2.16)$$

where mechanical stress F_i caused by thermal expansion during thermal cycling can be derived as

$$F_i = \frac{E_i S_i \Delta h_i}{h_i} \quad (2.17)$$

where E_i is Young's modulus of the material for the ith layer. Displacement Δh_i caused by thermal expansion every thermal cycle $\Delta T_a = T_{amax} - T_{amin}$ can be expressed as

$$\Delta h_i = \beta_i h_i \Delta T_a \quad (2.18)$$

where β_i is the CTE of the ith layer material. Therefore, the (2.16) can be rewritten as

$$W_i = \frac{F_i \Delta h_i}{V_i} = \frac{E_i S_i \Delta h_i^2}{S_i h_i^2} = E_i (\beta_i \Delta T_a)^2 \quad (2.19)$$

The thermal cycling lifetime involves thermo-mechanical coupling in the DSC power module. According to (2.15)–(2.19), the VAIED can be rewritten

$$W = D_W (\Delta T_a)^2 \quad (2.20)$$

where D_W is a coefficient for the VAIED, which is defined as

$$D_W = \frac{J}{V} = \frac{\sum_{i=1}^{10} E_i \beta_i^2 V_i}{\sum_{i=1}^{10} V_i} \tag{2.21}$$

where $J = \sum E_i \beta_i^2 V_i$ is a constant related to the property of the packaging material. $V = \sum V_i = S_i h_i = d_i^2 h_i$ is total volume related to the size of the packaging structure. In this chapter, the D_W is chosen as a mixed indicator to quantify the material properties and structure sizes, which can also be utilized to characterize the thermal cycling lifetime.

According to Tables 2.1 and 2.2, in the condition of $\Delta T_a = 80$ K, the von Mises stress of the DSC power module is illustrated by using the FEA approach. The topside and downside surfaces are attached to the water coolant in order to emulate the EV application conditions. As can be observed, during the thermal cycling caused by ambient temperature, the mismatched CTEs of the heterogeneous layers lead to an imbalance expansion of each layer. The internal mechanical stresses of the DSC power module further result in the extrusion and bend of layers.

In the same condition of Fig. 2.11, the breakdown of inelastic energy density is demonstrated in Fig. 2.12. As seen in (2.20) and (2.21), the height, CTE, and Young's Modulus of each layer comprehensively determine the mechanical stress of the DSC power module. As mentioned before, compared with the Mo spacer, the Cu spacer can reduce the thermal resistance of the DSC power module. However, according to (2.19) and Table 2.1, the inelastic energy density of the Cu spacer is 5.4 times of the Mo spacer. Additionally, the solder attachment and DBC Cu layers contribute nearly 90% of mechanical stresses, and they are weak points during thermal cycling. Compared with the solder, the influence of the DBC Cu is much more notable.

Besides, the FEA is implemented to ensure the proposed mechanical stress model. The von Mises stress of each layer achieved by the multi-physics tool COMSOL

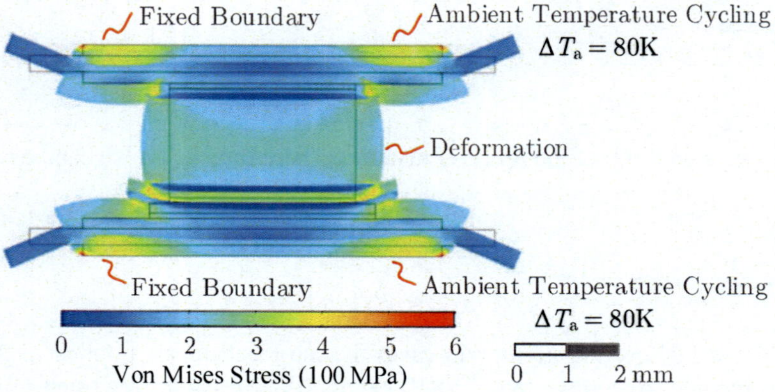

Fig. 2.11 Von Mises stress of DSC power module by using FEA

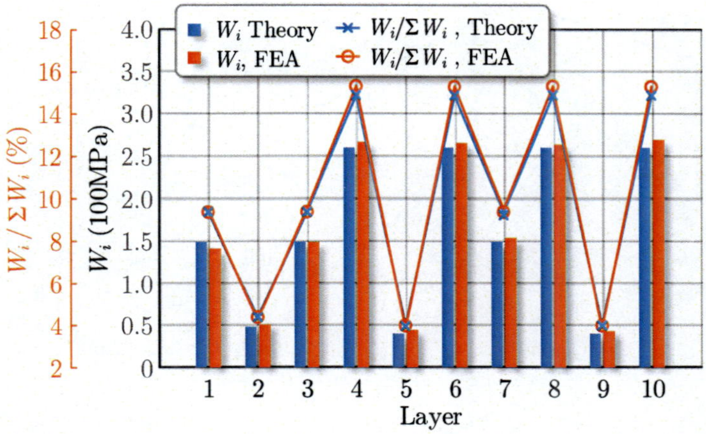

Fig. 2.12 Breakdown of the inelastic energy density of DSC power module

approximates the inelastic energy density modelled in (2.19). As seen, the proposed model is feasible to characterize the mechanical stress of the DSC power module.

As indicated in Fig. 2.12, the DBC metal and solder attachment mainly account for the mechanical stress of the DSC power module. How CTE and Young's Modulus of DBC metal and solder attachment influence the D_W is demonstrated in Fig. 2.13. Height, CTE, and Young's Modulus of these items comprehensively and nonlinearly determine the mechanical stress. Compared with the solder attachment, the influence of DBC metal is much more notable.

According to (2.21), the sensitivities of D_W to the height, Young's modulus, and CTE of each layer are respectively derived as

$$\begin{cases} \frac{\partial D_W}{\partial h_i} = \frac{E_i \beta_i^2 S_i V - S_i J}{V^2} \\ \frac{\partial D_W}{\partial E_i} = \frac{\beta_i^2 S_i h_i}{V} \\ \frac{\partial D_W}{\partial \beta_i} = \frac{2 E_i \beta_i S_i h_i}{V} \end{cases} \quad (2.22)$$

Sensitivities of D_W to structure heights and material properties are demonstrated in Fig. 2.14. Except for DBC ceramic layer, D_W increases with the height of the layer. Increasing Young's modulus of implemented material increases the D_W. Besides, the D_W is very sensitive to Young's modulus of the DBC metal. Increasing the CTE of used material also increases the D_W, while D_W is very sensitive to the CTE of DBC ceramic.

With (2.14) and (2.20), the sensitivity of thermal cycling lifetime to the D_W can be yielded

$$\frac{\partial N_t}{\partial D_W} = \frac{\partial N_t}{\partial W} \frac{\partial W}{\partial D_W} = \left(\lambda_1 \lambda_2 W^{\lambda_2 - 1} + \frac{\lambda_4 \theta_c}{\lambda_3} W^{\lambda_4 - 1} \right) (\Delta T_a)^2 \quad (2.23)$$

Therefore, the lifetime improvement by using reduced DW can be expressed as

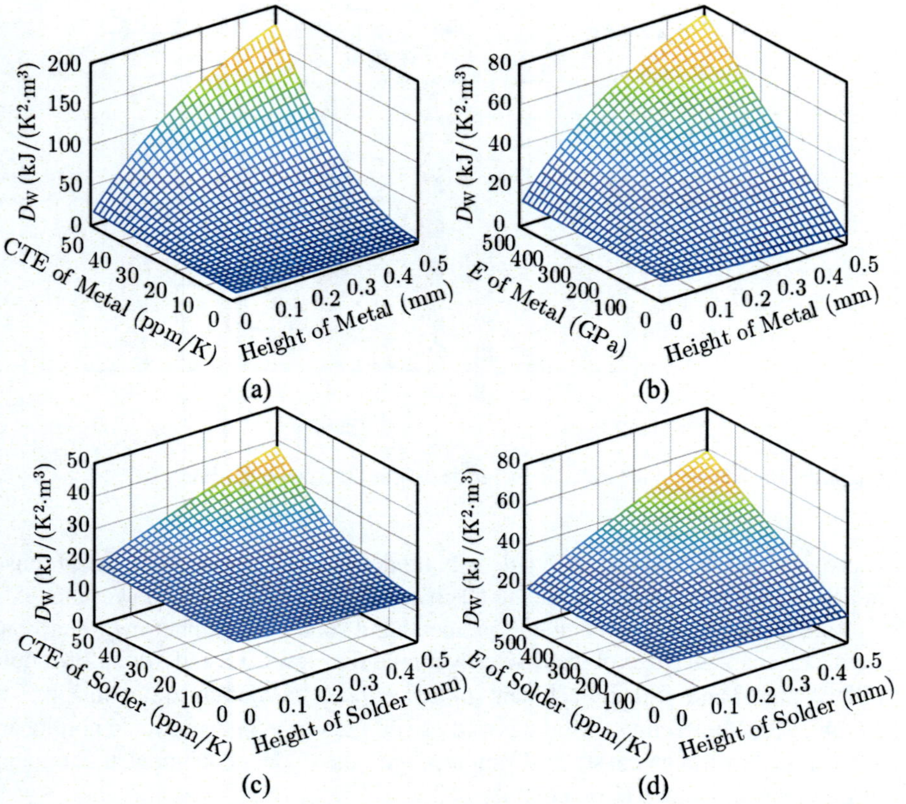

Fig. 2.13 Principles of inelastic energy density influenced by **a** CTE and height of DBC metal, **b** Young's modulus and height of DBC metal, **c** CTE and height of solder, and **d** Young's modulus and height of solder

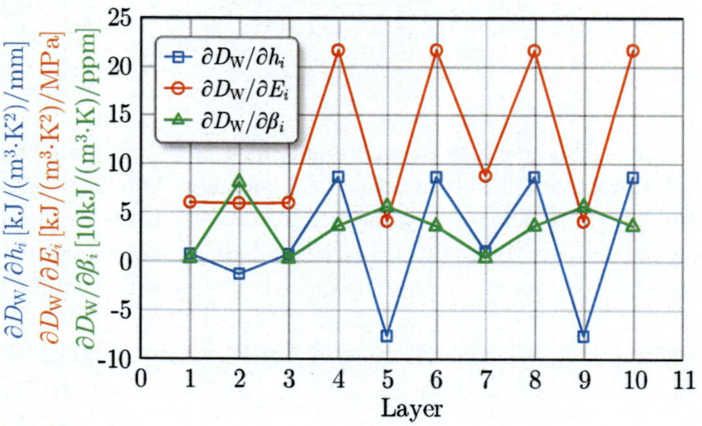

Fig. 2.14 Sensitivities of inelastic energy density to structure heights and material properties

$$\frac{\Delta N_t}{N_t} = \frac{1}{N_t}\frac{\partial N_t}{\partial D_W}\Delta D_W = \frac{\lambda_1\lambda_2 W^{\lambda_2-1} + \frac{\lambda_4\theta_c}{\lambda_3}W^{\lambda_4-1}}{\lambda_1 W^{\lambda_2} + \frac{\theta_c}{\lambda_3}W^{\lambda_4}}(\Delta T_a)^2\Delta D_W \tag{2.24}$$

As seen, owing to the decreased D_W by using optimal design, the extended thermal cycling lifetime of the DSC power module can be achieved. Moreover, the effectiveness of lifetime promotion is much more evident in the condition of deep thermal cycling with large ΔT_a.

2.3.3 The Proposed Multi-objective Optimization Model

As mentioned before, to minimize R_{th}, small and thin layers are required to enhance heat dissipation. However, to minimize D_W, wide and thick layers are desired to enlarge the volume of the DSC power module, in some conditions. According to (2.1) and (2.21), the D_W-R_{th} trajectory dominated by the heights of the DSC power module is demonstrated in Fig. 2.15. Except for the height of DBC metal layer h_8, the heights of DBC ceramic, solder, and spacer (h_5, h_7, and h_2) make the R_{th} and D_W cannot be simultaneously minimized. Obviously, there is a tradeoff between thermal resistance and mechanical stress to optimize the performance of the DSC power module.

As mentioned before, to minimize R_{th}, small and thin layers are required to enhance heat dissipation. However, to minimize D_W, wide and thick layers are desired to enlarge the volume of the DSC power module, in some conditions. According to (2.1) and (2.21), the D_W-R_{th} trajectory dominated by the heights of the DSC power module is demonstrated in Fig. 2.15. Except for the height of DBC metal layer h_8,

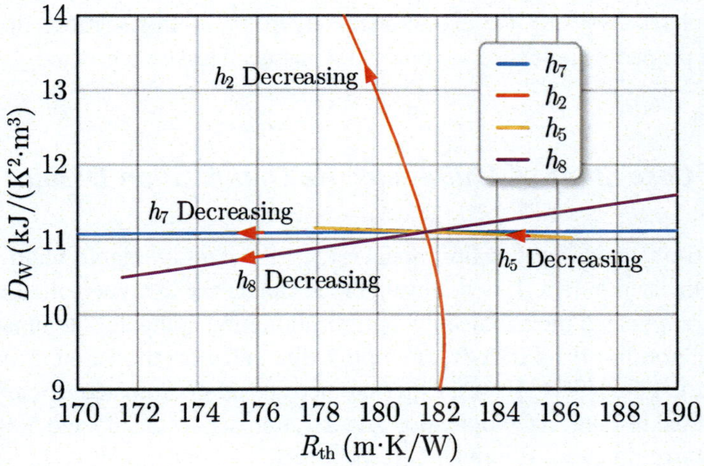

Fig. 2.15 Thermal resistance and inelastic energy density of DSC power module influenced by heights of different layers

Table 2.3 Boundaries of optimization of DSC power module

Variables	h_1	h_2	h_3	h_4	h_5	h_6	h_7	h_8	h_9	h_{10}	a_1	a_2
Min (mm)	0.08	2	0.08	0.1	0.10	0.1	0.08	0.1	0.1	0.1	1	$5\%d_c$
Max (mm)	0.20	10	0.20	0.4	0.63	0.4	0.20	0.4	0.63	0.4	10	$45\%d_c$

the heights of DBC ceramic, solder, and spacer (h_5, h_7, and h_2) make the R_{th} and D_W cannot be simultaneously minimized. Obviously, there is a tradeoff between thermal resistance and mechanical stress to optimize the performance of the DSC power module.

To coordinate the thermo-mechanical tradeoff of the DSC power module, a multi-objective optimization model is proposed, which can be expressed as

$$\begin{aligned} &\text{objective:} \quad \begin{cases} \min\ R_{th} \\ \min\ D_W \end{cases} \\ &\text{subject to:} \quad \begin{cases} h_{i\min} \le h_i \le h_{i\max} \\ a_{j\min} \le a_j \le a_{j\max} \end{cases} \end{aligned} \tag{2.25}$$

where $h_{i\min}$, $h_{i\max}$, $a_{j\min}$, and $a_{j\max}$ are boundaries limited by manufacturing technology, as listed in Table 2.3.

2.4 Solution of Proposed Thermo-mechanical-Coordinated Multi-objective Optimization Design

According to the multi-objective optimization model in (2.25), the Pareto solutions are calculated by using the evolutionary algorithm in this section. Besides, how material properties influence the optimal structure sizes are presented.

2.4.1 Case Study of Multi-objective Optimization Design

The multi-objective optimization model in (2.25) is a multi-variable, multi-constraint, and nonlinear problem. It is not available to obtain the analytic solution. To solve such a complicated multi-objective optimization model, the non-dominated sorting genetic algorithm II (NSGA-II) is employed to calculate the Pareto solutions. The flow-chart of the NSGA-II for the thermo-mechanical-coordinated automated design is illustrated in Fig. 2.16. Population and iteration are set as 300 and 500. Mutation and crossover rates are 0.3 and 0.1, respectively.

According to the parameters in Tables 2.1, 2.2 and 2.3, the NSGA-II algorithm in Fig. 2.16 is implemented to solve the multi-objective optimization design of the

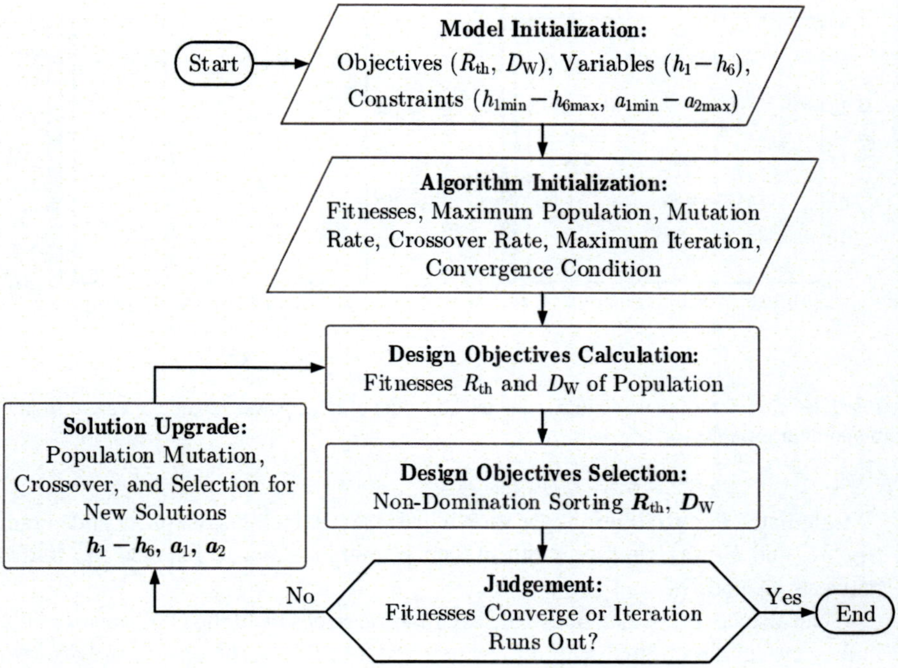

Fig. 2.16 Flowchart of automated multi-objective optimal design of DSC power module by using NSGA-II algorithm

DSC power module. The algorithm is individually activated ten times, and the corresponding convergence processes are demonstrated in Fig. 2.17. As seen, owing to the proper population and iteration opinions, the thermal and mechanical objectives simultaneously converge after 400 iterations.

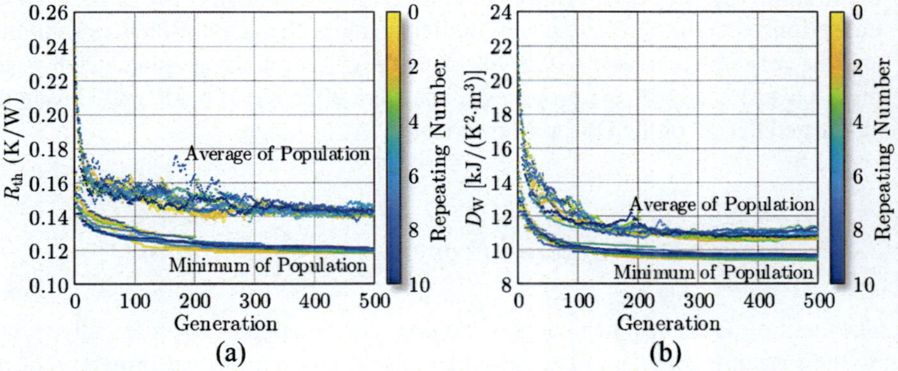

Fig. 2.17 Convergence of NSGA-II algorithm for design objectives **a** R_{th} and **b** D_W

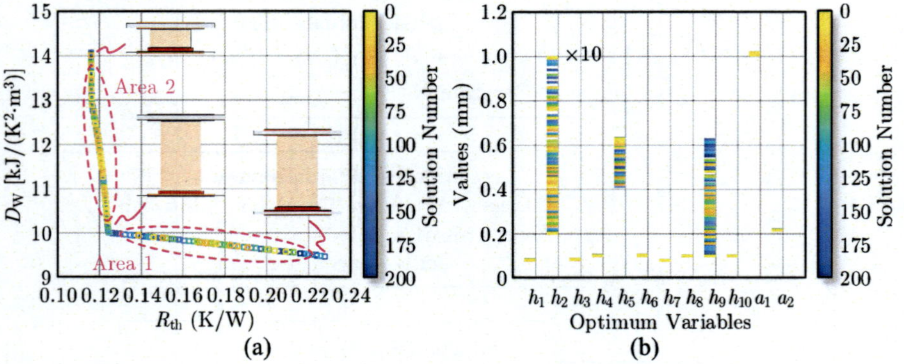

Fig. 2.18 Solution of multi-objective co-design model of DSC power module. **a** Pareto front and **b** optimum variables

Calculated Pareto solutions are shown in Fig. 2.18. Heights of spacer and ceramic (h_2, h_5, and h_9), as well as width a_1, are primary factors to balance the tradeoff between R_{th} and D_{W}.

It can also be found that, due to the asymmetric structure of the DSC power module in the vertical direction, the optimized heights of DBCs on the upside and downside are not the same. The Pareto solutions obviously consist of two areas. In area 1, the height of the DBC ceramic on the downside is the dominant factor to regulate the thermal resistance. In area 2, the height of the spacer is the main contributor to regulating mechanical stress.

The obtained Pareto solutions are confirmed by using the multi-physics coupling analysis. The achieved optimum packaging structure variables and values are sent to the multi-physics tool COMSOL. By implementing an FEA-based thermo-mechanical coupling analysis, the junction-case thermal resistance and mechanical stress of the DSC power module can be found. The multi-physics results are compared to the multi-objective optimization models in (2.25), as shown in Fig. 2.19. As seen, the optimized multi-objective results entirely agree with the multi-physics solutions, and the relative errors are within 5%, generally. Therefore, as guaranteed by the multi-physics analysis, the proposed models are effective to guide and elevate the automated design of the DSC power module.

2.4.2 Influence of Material Properties on Optimization

The thermo-mechanical behaviors of the DSC power module are profoundly shaped by the structure sizes and material properties. The optimized performances of the DSC power module influenced by different materials are addressed in this subsection.

Considering different attachment solders, optimal solutions are obtained in Fig. 2.20. Heights of the spacer and downside DBC ceramic (h_2 and h_9) are the

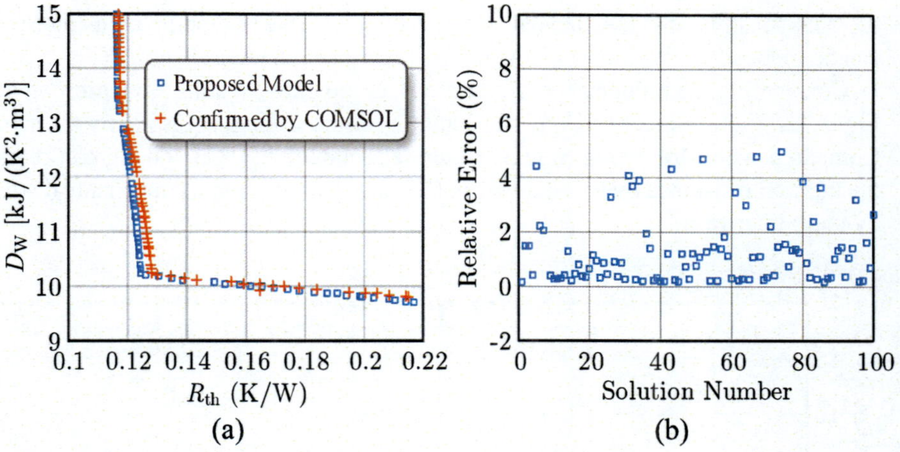

Fig. 2.19 Comparison of multi-objective optimization model and multi-physics coupling analysis. **a** Pareto solutions and **b** relative errors

most useful variables to balance the thermo-mechanical feature. Compared with SAC or SnPb alloys, the thermal resistance of DSC the power module by using Ag paste can be decreased by 46%. By using Ag paste, smaller thermal resistance and mechanical stress can be achieved to improve the lifetime of the DSC power module.

By implementing different ceramic materials for DBC, the solutions are compared, as demonstrated in Fig. 2.21. As can be seen, AlN benefits from lower thermal resistance, while Si_3N_4 can achieve lower mechanical stress. Compared with Al_2O_3, they can provide much better thermo-mechanical performance. Compared to

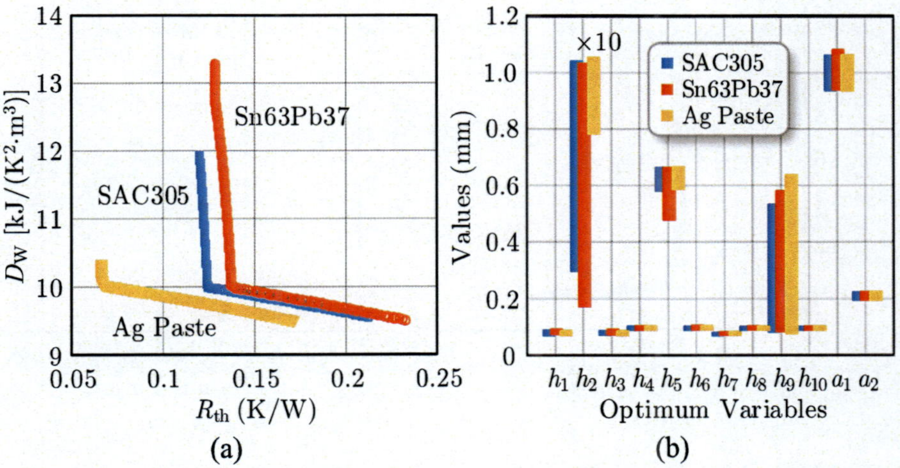

Fig. 2.20 Optimal packaging structure solution affected by attachment solders. **a** Pareto front **b** optimum variables

the AlN substrate, the mechanical stress of the DSC power module by using Si_3N_4 can be reduced by 8%; however, its thermal resistance increases by 21%.

Considering the influence of spacer metal, comparative results are represented in Fig. 2.22. Thermo-mechanical performance by using Cu and Ag is nearly the same. Compared with Mo, owing to higher thermal conductivity, Cu and Ag can reduce the thermal resistance by less than 10%. However, Mo can reduce mechanical stress by more than 18%.

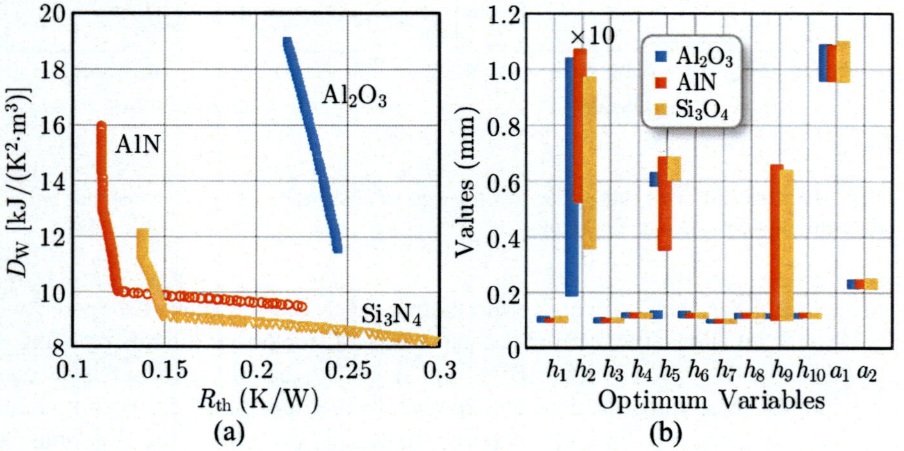

Fig. 2.21 Optimal packaging structure solution affected by DBC ceramics. **a** Pareto front **b** optimum variables

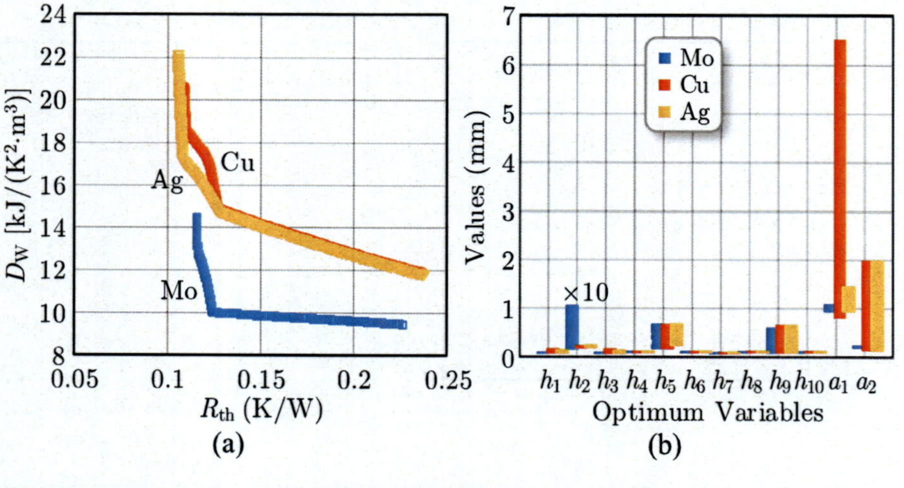

Fig. 2.22 Optimal packaging structure solution affected by spacer metals. **a** Pareto front **b** optimum variables

In summary, there is an apparent tradeoff between thermal resistance and mechanical stress to design the DSC power module. To balance and promote the thermo-mechanical performances, ceramic AlN for the DBC substrate, Ag paste for the attachment solder, and metal Mo for the spacer are recommended.

2.5 Experimental Results

To ensure the feasibility of the design methodology of the DSC power module, the realization of the designed DSC power module is demonstrated in this Section.

According to the optimal design results in Sect. 2.4, a single-switch DSC power module prototype is fabricated, as depicted in Fig. 2.23. SiC MOSFET bare die H1M120M060 from Hestia is employed. To achieve a solderable chip for DSC application, metal Ag is deposited on the top source and gate pads of the chip. Similarly, to guarantee the Mo spacer is solderable, a thin Ti–Ni–Ag film is electroplated on the spacer. AlN ceramic substrates and SAC305 solders are utilized in the DSC power module prototype. Cu terminal pins are implemented as the lead-frame of the power module.

The step-by-step fabrication procedures of the DSC power module are illustrated in Fig. 2.24. From bottom to top, the heterogeneous layers are stacked one-by-one. Three solder layers are employed to assemble the DSC power module by using reflow soldering. The bonding wire is used to connect the gate pad.

With the aid of the double-pulse test, the transient behaviors of the DSC power module in the condition of different gate resistances are observed, as can be seen in Fig. 2.25. As seen, small gate resistance can elevate the switching speed and reduce the switching loss of the switch.

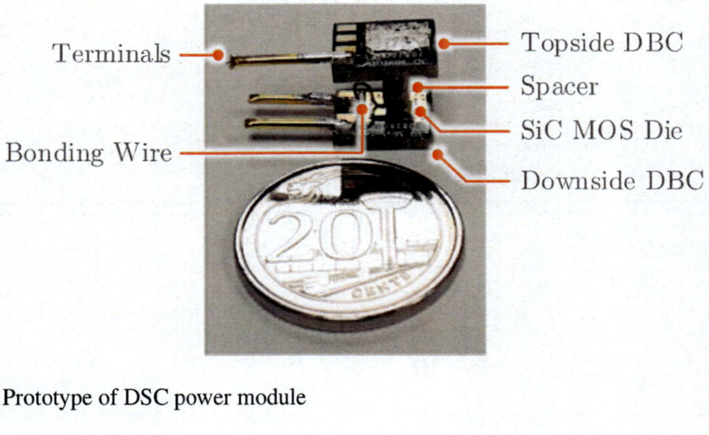

Fig. 2.23 Prototype of DSC power module

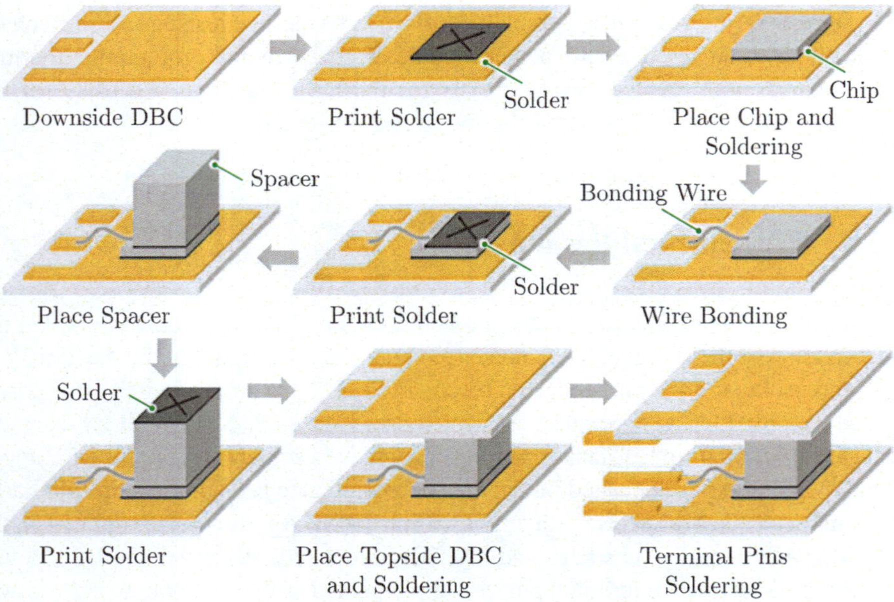

Fig. 2.24 Fabrication procedure of DSC power module

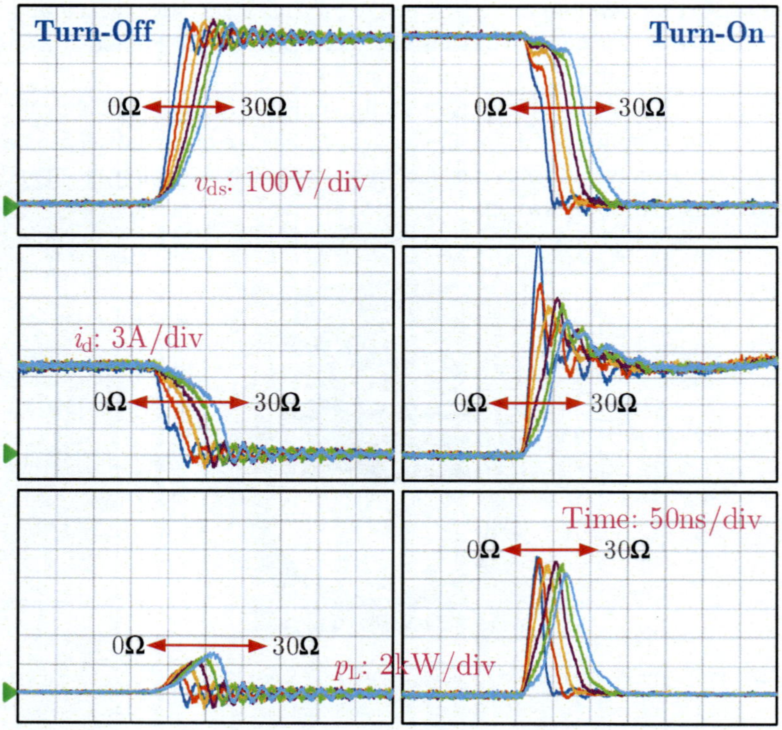

Fig. 2.25 Experimental results of fabricated DSC power module

2.6 Conclusions

DSC power module is a traditional powertrain solution for EV deployment that improves the high lifetime expectancy and long-term reliability requirements. Without mathematical model guidelines, the conventional DSC power module design is a trial-and-error, expensive, and time-consuming process. This chapter proposes a multi-objective-oriented and thermo-mechanical-coordinated design methodology to tackle the under-optimization issue and improve the performance of the emerging DSC power module. The multi-physics FEA tool COMSOL is used to develop and validate the mathematical models of thermal resistance and mechanical stress of the DSC power module. A multi-objective optimization model is proposed to balance the tradeoff between thermal and mechanical performances of the DSC power module. These proposed models are also used to guide the optimization of the DSC power module's material properties and structural sizes. To address the multi-objective optimization problem, the NSGA-II algorithm is used. The heights of spacer metal on the upside and DBC ceramic on the downside are the main factors to balance the thermal and mechanical characteristics, according to the optimal results verified by the FEA tool. Furthermore, ceramic AlN, metal Mo, and solder Ag paste are suggested as more suitable materials for the DSC power module to improve thermal–mechanical properties. Prototype and experiments of a DSC power module ensure the proposed design methodology. The proposed multi-objective optimization design methodology is a novel procedure for the next-generation DSC power module, with the goals of improving design efficiency, lowering design risk, and balancing the trade-off between thermal resistance and mechanical stress.

References

1. A. Emadi, S. S. Williamson, and A. Khaligh, "Power electronics intensive solutions for advanced electric, hybrid electric, and fuel cell vehicular power systems," *IEEE Trans. Power Electron.*, vol. 21, no. 3, pp. 567–577, 2006.
2. Yole Développement. *EV-HEV market and technology trends.* [Online]. www.psma.com, 2015.
3. D. Hirschmann, D. Tissen, S. Schroder, and R. W. D. Doncker, "Reliability prediction for inverters in hybrid electrical vehicles," *IEEE Trans. Power Electron.*, vol. 22, no. 6, pp. 2511–2517, 2007.
4. B. Ji, X. Song, W. Cao, V. Pickert, Y. Hu, J. W. Mackersie, and G. Pierce, "In situ diagnostics and prognostics of solder fatigue in IGBT modules for electric vehicle drives," *IEEE Trans. Power Electron.*, vol. 30, no. 3, pp. 1535–1543, 2015.
5. Z. Liang, P. Ning, F. Wang, and L. Marlino, "A phase-leg power module packaged with optimized planar interconnections and integrated double-sided cooling," *IEEE J. Emerg. Sel. Topics Power Electron.*, vol. 2, no. 3, pp. 443–450, Sep. 2014.
6. S. Seal and H. A. Mantooth, "High performance silicon carbide power packaging—Past trends, present practices, and future directions", *Energies*, vol. 10, no. 3, p. 341, 2017.
7. H. Zhang, S. S. Ang, H. A. Mantooth, and S. Krishnamurthy, "A high temperature, double-sided cooling SiC power electronics module," in *IEEE ECCE*, 2013, pp. 2877–2883.
8. N. Zhu, H. A. Mantooth, D. Xu, M. Chen, and M. D. Glover, "A solution to press-pack packaging of SiC MOSFETs," *IEEE Trans. Ind. Electron.*, vol. 64, no. 10, pp. 8224–8234, 2017.

9. J. O. Gonzalez, O. Alatise, A. M. Aliyu, P. Rajaguru, A. Castellazzi, L. Ran, P. A. Mawby, and C. Bailey, "Evaluation of SiC schottky diodes using pressure contacts," *IEEE Trans. Ind. Electron.*, vol. 64, no. 10, pp. 8213–8223, 2017.
10. S. Kicin, M. Laitinen, C. Haederli, J. Sikanen, R. Grinberg, C. Liu, J. H. Fabian, and A. Hamidi, "Low-voltage AC drive based on double-sided cooled IGBT press-pack modules," *IEEE Trans. Ind. Appl.*, vol. 48, no. 6, pp. 2140–2146, 2012.
11. P. Ning, Z. Liang, and F. Wang, "Power module and cooling system thermal performance evaluation for HEV application," *IEEE J. Emerg. Sel. Topics Power Electron.*, vol. 2, no. 3, pp. 487–495, 2014.
12. Z. Liang, P. Ning, and F. Wang, "Development of advanced all-SiC power modules," *IEEE Trans. Power Electron.*, vol. 29, no. 5, pp. 2289–2295, 2014.
13. S. Li, L. M. Tolbert, F, and F. Z. Peng, "Stray inductance reduction of commutation loop in the P-cell and N-cell-based IGBT phase leg module," *IEEE Trans. Power Electron.*, vol. 29, no. 7, pp. 3616–3624, 2014.
14. F. Yang, Z. Wang, Z. Liang, and F. Wang, "Electrical performance advancement in SiC power module package design with kelvin drain connection and low parasitic inductance," *IEEE J. Emerg. Sel. Topics Power Electron.*, vol. 7, no. 1, pp. 84–98, 2019.
15. L. Yin, C. Kapusta, A. Gowda, and K. Nagarkar, "A wire-bondless packaging platform for silicon carbide power semiconductor devices," *Journal of Electronic Packaging*, vol. 140, no. 3, pp. 031009–1–8.
16. E. Vagnon, P. O. Jeannin, J. C. Crebier, and Y. Avenas, "A bus-bar-like power module based on three-dimensional power-chip-on-chip hybrid integration," *IEEE Trans. Ind. Appl.*, vol. 46, no. 5, pp. 2046–2055, 2010.
17. G. Regnat, P. O. Jeannin, D. Frey, J. Ewanchuk, S. V. Mollov, and J. P. Ferrieux, "Optimized power modules for silicon carbide MOSFET," *IEEE Trans. Ind. Appl.*, vol. 54, no. 2, pp. 1634–1644, 2018.
18. X. Zhao, B. Gao, L. Zhang, D. C. Hopkins, and A. Q. Huang, "Performance optimization of A 1.2kV SiC high density half bridge power module in 3D package", in *IEEE APEC*, 2018, pp. 1266–1271.
19. P. Lasserre, D. Lambert, and A. Castellazzi, "Integrated Bi-directional SiC MOSFET power switches for efficient, power dense and reliable matrix converter assembly," in *IEEE WiPDA*, 2016, pp. 188–193.
20. M. Wang, Y. Mei, X. Li, R. Burgos, D. Boroyevich, and G. Q. Lu, "Pressureless silver sintering on nickel for power module packaging," *IEEE Trans. Power Electron.*, vol. PP, no. 99, pp. 1–1, 2019.
21. Y. H. Mei, J. Y. Lian, X. Chen, G. Chen, X. Li, and G. Q. Lu, "Thermo-mechanical reliability of double-sided IGBT assembly bonded by sintered nanosilver," *IEEE Trans. Device and Mater. Rel.*, vol. 14, no. 1, pp. 194–202, 2014.
22. K. Sienski, R. Eden, and D. Schaefer, "3-D electronic interconnect packaging," in *IEEE Aerospace Applications Conference*, 1996, pp. 363–373.
23. H. Tanisawa, F. Kato, K. Koui, S. Sato, K. Watanabe, H. Takahashi, Y. Murakami, and H. Sato, "Transient thermal characteristics of high-temperature SiC power module enhanced with Al-bump technology," *Japanese Journal of Applied Physics*, vol. 57, 2018, pp. 04FR10–1–5.
24. S. Zhu, Y. Li, Y. Wang, Y. Ma, C. Wu, M. Jiao, Z. Zhao, and J. Yu, "Advanced double sided cooling IGBT module and power control unit development," in *IEEE IWIPP*, 2017, pp. 1–4.
25. A. Matsushita, R. Saito, T. Tokuyama, K. Nakatsu, and T. Kimura, "An experimental study on the thermal performance of double-side direct-cooling power module structure," in *PCIM Europe*, 2016, pp. 1–5.
26. I. Aranzabal, I. M. de Alegria, N. Delmonte, P. Cova, and I. Kortabarria, "Comparison of the heat transfer capabilities of conventional single-phase and two-phase cooling systems for electric vehicle IGBT power module," *IEEE Trans. Power Electron.*, vol. PP, no. 99, pp. 1–1, 2018. (Early Access)
27. C. Buttay, J. Rashid, C. M. Johnson, P. Ireland, F. Udrea, G. Amaratunga, and R. K. Malhan, "High performance cooling system for automotive inverters," in *IEEE European Conference on Power Electronics and Applications*, 2007, pp. 1–9.

28. D. R. M. Woo, H. H. Yuan, J. A. J. Li, L. J. Bum, and Z. Hengyun, "Miniaturized double side cooling packaging for high power 3 phase SiC inverter module with junction temperature over 220°C," in *IEEE ECTC*, 2016, pp. 1190–1196.
29. S. Dwarakanath, P. M. Raj, V. Smet, V. Sundaram, M. D. Losego, and R. Tummala, "High-temperature and moisture-ageing reliability of high-density power packages for electric vehicles," in *IEEE ECTC*, 2018, pp. 179–184.
30. A. Grassmann, O. Geitner, W. Hable, C. Neugirg, A. Schwarz, F. Winter, and I. Yoo, "Double sided cooled module concept for high power density in HEV applications," in *IEEE PCIM*, 2015, pp. 1–7.
31. A. K. Solomona, R. Skuriat, A. Castellazzi, and P. Wheeler, "Modular integration of a matrix converter," *IEEJ Trans. Electrical and Electronic Engineering*, vol. 11, pp. 103–111, 2016.
32. C. J. Marbut, M. Montazeri, and D. R. Huitink, "Rapid solder interconnect fatigue life test methodology for predicting thermomechanical reliability," *IEEE Trans. Device and Mater. Rel.*, vol. 18, no.3, pp. 412–421, 2018.
33. T. M. Evans, Q. Le, S. Mukherjee, I. A. Razi, T. Vrotsos, Y. Peng, and A. H. Mantooth, "PowerSynth: A power module layout generation tool", *IEEE Trans. Power Electron.*, vol. PP, no. 99, pp. 1–1, 2018.
34. H. A. Mantooth, T. Evans, C. Farnell, Q. Le, and R. Murphree, "Emerging trends in silicon carbide power electronics design", *CPSS Trans. Power Electron. Appl.*, vol. 2, no. 3, pp. 161–169, 2017.
35. T. Dragicevic, P. Wheeler, and F. Blaabjerg, "Artificial intelligence aided automated design for reliability of power electronic systems," *IEEE Trans. Power Electron.*, vol. PP, no. 99, pp. 1–1, 2018. (Early Access)
36. A. Bindra and A. Mantooth, "Modern tool limitations in design automation: Advancing automation in design tools is gathering momentum," *IEEE Power Electron. Mag.*, vol. 6, no. 1, pp. 28–33, 2019.
37. Yole Développement. *Infineon FF400R07A01E3.* [Online]. www.systemplus.fr, 2018.
38. Infineon. FF400R07A01E3_S6. [Online]. www.infineon.com, 2018.
39. A. Hanif, Y. Yu, D. Devoto, and F. H. Khan, "A comprehensive review toward the state-of-the-art in failure and lifetime predictions of power electronic devices", *IEEE Trans. Power Electron.*, vol. PP, no. 99, pp. 1–1, 2018.
40. B. Hu, J. O. Gonzalez, L. Ran, H. Ren, Z. Zeng, W. Lai, B. Gao, O. Alatise, H. Lu, C. Bailey, and P. Mawby, "Failure and reliability analysis of a SiC power module based on stress comparison to a Si device", *IEEE Trans. Device and Mater. Rel.*, vol. 17, no. 4, pp. 727–737, 2017.
41. R. Darveaux, "Effect of simulation methodology on solder joint crack growth correlations", in *Proc. of IEEE ECTC*, pp. 1048–1058, 2000.

Chapter 3
The Proposed Multi-objective Design of Output LC Filter for Buck Converter via the Coevolving-AMOSA Algorithm

3.1 Introduction

Both industry and daily life rely on buck converters in significant ways. In industry, Buck converters are applied in electric vehicles [1], renewable energy systems [2], and others [3] for power regulation and voltage conversion [4]. In our daily life, the applications of Buck converters are everywhere, such as portable electronic devices [5], power audio systems [6], photovoltaic systems [7], etc.

Passive output *LC* filters are often used to decrease the ripples produced by Buck converters since they are inexpensive and simple to construct. Traditional design [8] of output *LC* filter mostly depends on the required output voltage and current ripple. However, the output *LC* filter not only affects output voltage and current ripples but also affects other performance [9] of Buck converters. For instance, the *LC* filter parameter choice will have a direct impact on the power loss of the Buck converter, which must be as little as possible to ensure good power efficiency. Inductance and capacitance levels will also affect the *LC* filter's volume. Additionally, the cut-off frequency must be low, which is also influenced by the output *LC* filter's characteristics, to guarantee greater filtering capabilities.

It appears that certain particular applications have rigorous guidelines for specific design goals. As described in Fig. 3.1, airplanes, satellites and electric vehicles demand high-efficiency products [10]. Battery adapters, rooftop PV, digital cameras, and LED, which have limited space, prefer more compact electronic devices [11]. Additionally, the Buck converter's *LC* filter with a lower cut-off frequency exhibits higher filtering performance, making it suited for audio amplifiers or MP3 players with stringent ripple reduction requirements [6]. Even while some applications have severe criteria for specific performance indicators, the Buck converter is still expected to perform at its best overall, therefore other design goals should also be taken into account.

However, the review of the literature indicates a large number of research articles on the design of *LC* filters that only discuss one design purpose, such as cost [12], volume [13], voltage quality [14], and reliability [15]. For instance, the volume of the

X. Zhang et al., *Automated Design of Electrical Converters with Advanced AI Algorithms*,
https://doi.org/10.1007/978-981-99-0459-4_3

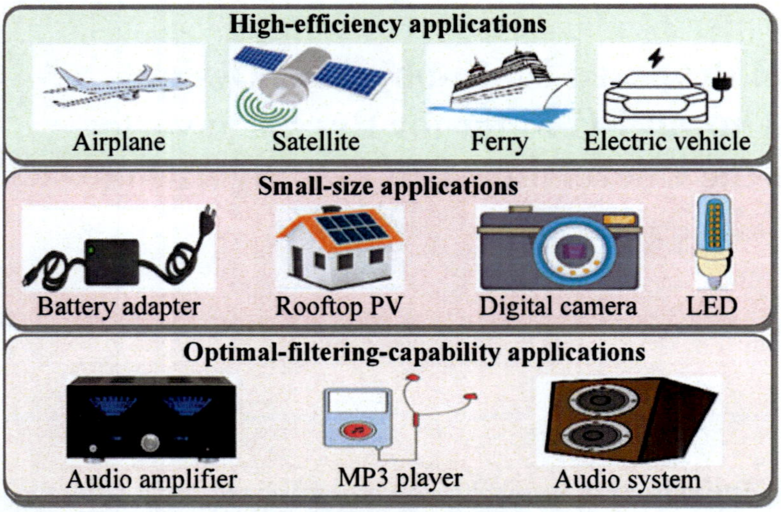

Fig. 3.1 Descriptions of requirements of different applications on Buck converters

capacitor is set as the design objective in [13]. To improve power efficiency, power loss is optimized in [16]. For optimal filtering performance, the cut-off frequency is considered in [17].

The trade-off interactions between many design objectives are what make multi-objective design for the *LC* filter in the Buck converter challenging. For instance, a smaller volume necessitates smaller inductance and capacitance values, which results in greater power loss and worse filtering performance [16]. It is acknowledged that certain academics are engaged in multi-objective *LC* filter Buck converter optimization study such as [9, 14, 18]. However, despite its importance for energy conservation and environmental friendliness, power efficiency is not taken into consideration in these study studies. **The first task of this chapter is to carry out the multi-objective design while taking the power efficiency, filtering capabilities, and volume of the *LC* filter in the Buck converter into consideration**.

In order to find the Pareto-Frontier, which is made up of all optimum solutions, multi-objective optimization methods are typically used to address multi-objective design issues. Three basic categories of multi-objective algorithms include population-based algorithms, decomposition-based algorithms, and indicator-based algorithms. The Pareto-Frontier is obtained via indicator-based algorithms using indicator functions. With decomposition-based algorithms, the original multi-objective issue is divided into many single-objective problems, which are then solved to provide the Pareto-Frontier. Multiple solutions (which make up the population) are evaluated simultaneously using population-based multi-objective algorithms, which can swiftly find the Pareto-Frontier. The commonly used multi-objective algorithms include NSGA-II [19], MOPSO [20], AMOSA [21], MOEA/D [22], and IBEA [22]. However, the non-uniform or partial Pareto-Frontier coverage of the current multi-objective algorithms substantially undermines the design's correctness [23]. **This**

chapter's second task is to perform the multi-objective design for the *LC* filter in the Buck converter while enhancing the uniformity and completeness of the Pareto-Frontier for more precise and fully optimal designs.

In order to create a completely optimized *LC* filter, a multi-objective design strategy for the output *LC* filter in the Buck converter is proposed in this chapter. In Stage 1, three design goals—power efficiency, cut-off frequency, and volume—are examined. Additionally, Stage 2 adopts the particularly proposed coevolving-AMOSA algorithm for this multi-objective design method to precisely discover the Pareto-Frontier. The final design options can then be chosen in Stage 3 along the determined Pareto-Frontier following the application requirements. Three design cases that correspond to three distinct application scenarios will be shown as design examples in this chapter.

The remainder of this chapter is set up as follows. Concerning the trade-off interactions between the three design objectives and the nonuniform and imperfect Pareto-Frontier achieved by popular multi-objective algorithms, problem descriptions will be given in Sect. 3.2. Section 3.3 will provide a thorough study of the three design goals, power efficiency, cut-off frequency, and volume. In Sect. 3.4, a detailed introduction to the proposed multi-objective design strategy for the *LC* filter in the Buck converter using the coevolving-AMOSA algorithm is provided. In Sect. 3.5, three design examples are listed, and in Sect. 3.6, the relevant experimental validation is provided. A summary of the conclusion follows.

3.2 Problem Descriptions for the Multi-objective Design of the Output LC Filter in Buck Converter

3.2.1 Preliminaries: Introduction to Pareto-Frontier

Since the optimization of certain objectives would result in the sacrifice of other objectives, it is difficult to arrive at a single global optimal design when numerous competing purposes are taken into account. As a result, a solution is said to be Pareto optimal if no adjustment could make all of the objectives better [24]. Additionally, all Pareto optima for this multi-objective optimization problem are contained in Pareto-Frontier. There isn't a design that can achieve more goals than those on the Pareto-Frontier. Figure 3.2 provides a Pareto-Frontier example for the minimization of f_1 and f_2, where f_1 and f_2 are negatively correlated.

In other words, Pareto-Frontier offers the best designs when goals are in trade-off relationships, and it can be used to create the best multi-objective designs for *LC* filters. One final design along the Pareto-Frontier might be chosen by taking into account various application needs.

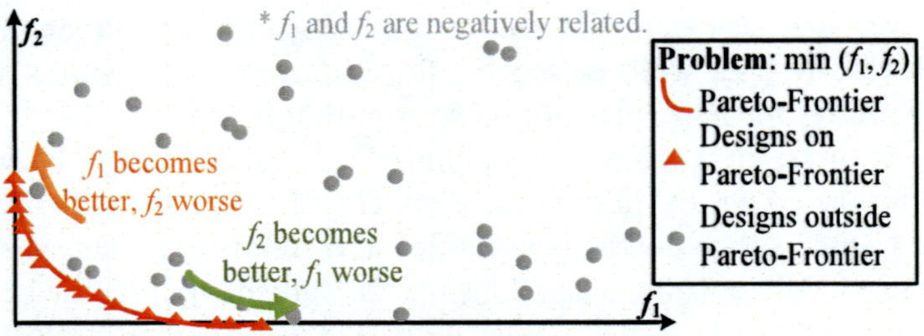

Fig. 3.2 Example of Pareto-frontier for the minimization of f_1 and f_2

3.2.2 Problem I: Trade-Off Relationships Among the Three Design Objectives for the Output LC Filter in Buck Converter

For the optimization of the output *LC* filter in the Buck converter, more complete design objectives should be taken into account to provide a holistic performance. This chapter considers power efficiency, cut-off frequency, and volume to create a small *LC* filter with enhanced efficiency and filtering capabilities for Buck converters.

There are trade-off relationships between these three objectives, though, as shown in Fig. 3.3. Smaller volume will result in lower efficiency and filtering capabilities since minimizing power loss is incompatible with minimizing volume, and minimizing volume is incompatible with minimizing cut-off frequency.

In order to create a completely optimized output *LC* filter for Buck converters, the first task of this chapter is to resolve the contradictory relationships among the three design objectives (power loss, cut-off frequency, and volume).

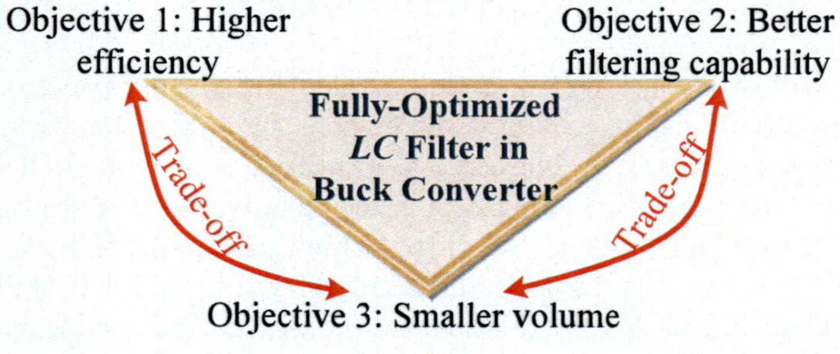

Fig. 3.3 Descriptions of the multi-objective design for the *LC* filter in Buck converter in this chapter

3.2.3 Problem II: The Nonuniform and Incomplete Coverage of Pareto-Frontier

The Pareto-Frontier method, which was described in Sect. 3.2.1, is typically used to achieve the optimization of several competing design objectives. You can also use multi-objective optimization techniques to find the Pareto-Frontier. The Pareto-Frontier, however, suffers from the following two flaws in the generally employed multi-objective optimization techniques.

The Pareto-non-uniform Frontier's coverage, as seen in Fig. 3.4, is the first downside. One or more regions are left empty because the derived Pareto-Frontier does not completely cover the space. The final acquired design solution will diverge from the required design solution if the desired *LC* filter design is not present, which will reduce the design accuracy.

The Pareto-insufficient Frontier's coverage, as seen in Fig. 3.5, is the second flaw. In this scenario, the Pareto-Frontier produced by the multi-objective optimization techniques in use has inadequate and insufficient coverage. As a result, the achieved design solution is probably not entirely optimal, which has a detrimental impact on the planned *LC* filter's performance.

To achieve an accurate and fully optimal design for the output *LC* filter in the Buck converter, it is envisaged that both the nonuniform and partial coverage shortcomings of the Pareto-Frontier outlined above will be avoided. As a result, in this chapter second task is to increase Pareto-uniformity Frontier and completeness.

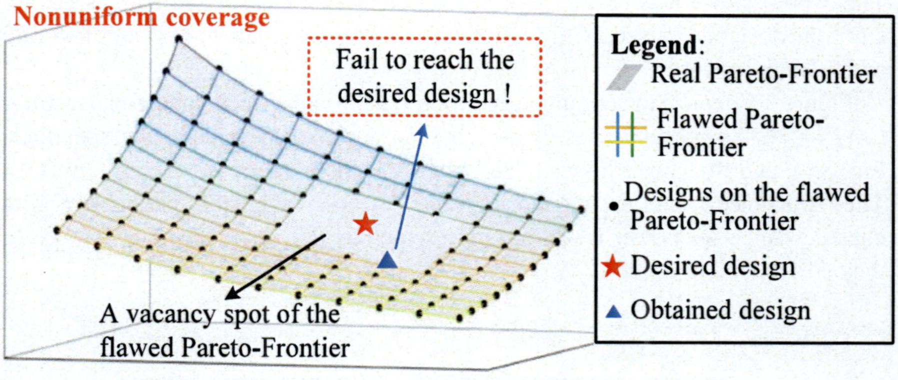

Fig. 3.4 Problem description: nonuniform coverage of Pareto-frontier

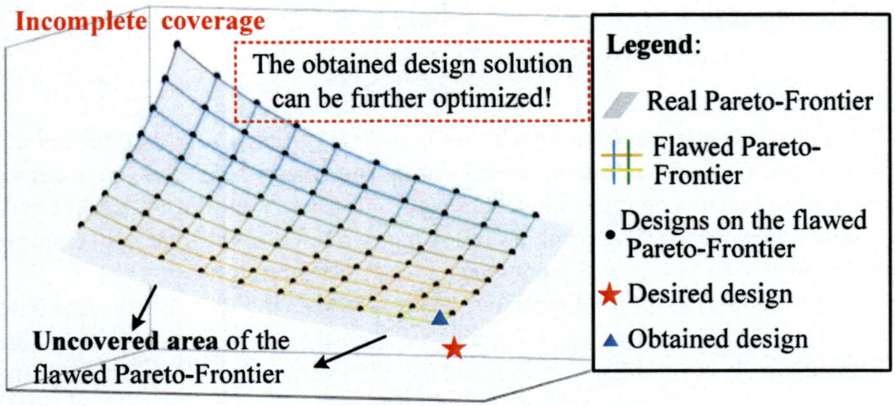

Fig. 3.5 Problem description: incomplete coverage of Pareto-frontier

3.3 Analysis of the Three Design Objectives for LC Filter: Power Efficiency, Cut-Off Frequency and Volume

3.3.1 Analysis of Design Objective 1: Optimized Total Power Loss for the Buck Converter of Optimal Power Efficiency

The overall power loss of the Buck converter is specified as the initial design target since the designed output *LC* filter will have an impact on the total power loss of the entire Buck converter.

Figure 3.6 depicts the circuit design for a synchronous Buck converter. According to [25–27], the power losses of switches S_L and S_H (S_L and S_H are specified in Fig. 3.6), and the power losses of *LC* filters are included in the overall power loss. The inductance *L* and capacitance *C* values are the designable parameters in this chapter, hence the power losses are represented as follows in terms of how closely they correspond to *L* and *C*.

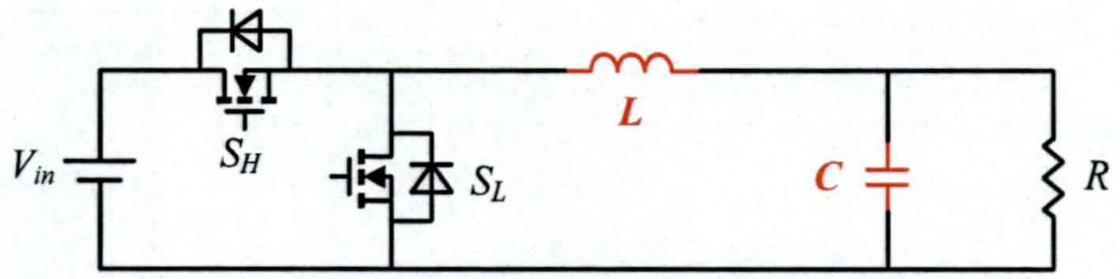

Fig. 3.6 Circuit diagram of the synchronous Buck converter

3.3.1.1 Driving Loss of Switches S_L and S_H (P_{l_dr})

Driving loss P_{l_dr} [27] is defined in (3.1).

$$P_{l\_dr} = 2Q_G V_{gs} f \tag{3.1}$$

where Q_G is the total gate charge of main switches, V_{gs} is the gate-to-source voltage and f is the switching frequency. P_{l_dr} is constant since Q_G, V_{gs} and f are fixed design specifications.

3.3.1.2 Conduction Loss of Switches S_L and S_H (P_{l_on} (L))

With [27], the conduction loss of S_L and S_H, is expressed as:

$$P_{l\_on}(L) = DR_{on}(\frac{1}{12}(\frac{V_{in} - V_o}{Lf}D)^2 + I_o^2) \tag{3.2}$$

where V_{in}, V_o are the input and output voltages. I_o is the output current. D is the duty cycle of high-side switch S_H. R_{on} is the equivalent drain-source on the resistance of switches. P_{l_on} relates to L, according to (3.2), in which V_o, I_o, D, R_{on} are constant design specifications.

3.3.1.3 Switching Loss of Switches S_L and S_H (P_{l_s})

The switching loss P_{l_s} of switches S_H and S_L is in (3.3) [27]:

$$P_{l\_s} = 0.5V_{in}I_L\left(t_{r\_H} + t_{f\_H}\right)f + 0.5V_{SD}I_L\left(t_{r\_L} + t_{f\_L}\right)f \tag{3.3}$$

where t_{r_H} and t_{f_H} are the rising and falling time of the high side switch S_H. t_{r_L} and t_{f_L} are the rising and falling time of the low side switch S_L. V_{SD} is the conduction voltage across the diode of S_L. As can be seen from (3.3), P_{l_s} is constant.

3.3.1.4 Core Loss of Inductor (P_{l_Fe} (L))

According to Reinert et al. [26], the core loss of inductor in Buck converter is computed by the Steinmetz equation.

$$P_{l\_Fe}(L) = k\Delta B^{\beta} f^{\alpha}\left[(D)^{1-\alpha} + (1-D)^{1-\alpha}\right]Vol_L(L) \tag{3.4}$$

where k, α, and β are the parameters in the Steinmetz equation and are constants when the inductor core material is selected. ΔB is the magnetic fluctuation, calculated by

Reinert et al. [26] and datasheets of inductor cores. From (3.4), P_{l_Fe} only relates to inductance L.

3.3.1.5 Copper Loss of Inductor (P_{l_Cu} (*L*))

The calculation of copper loss of inductor [25] is in (3.5):

$$P_{l\_Cu}(L) = (R_{dc} + R_{ac})(\frac{1}{12}(\frac{V_{in} - V_o}{Lf}D)^2 + I_o^2) \tag{3.5}$$

where R_{dc} is the dc winding resistance of the inductor and is obtained from inductor datasheets, and R_{ac} is the ac winding resistance considering skin and proximity effects and is computed with [25]. From (3.5), P_{l_Cu} is related to L only.

3.3.1.6 Power Loss of Capacitor (P_{l_C} (*L*, *C*))

The loss of capacitor is computed by (3.6), where I_{Ck} is the root mean square of the *kth* harmonic current on the capacitor, and is related to both L and C [18]. tanδ is constant and can be found from the datasheets of capacitors. According to (3.6), with the increasing values of L and C, P_{l_C} decreases.

$$P_{l\_C}(L, C) = \sum_{k}^{\infty} I_{Ck}^2 \cdot \frac{\tan\delta}{2\pi kfC} \tag{3.6}$$

3.3.1.7 Total Power Loss (P_{l_tot} (*L*, *C*))

The total power loss sums (3.1)–(3.6) together, as shown in (3.7). According to (3.7) and Fig. 3.7, P_{l_tot} relates to both inductance L and capacitance C.

$$\begin{aligned} P_{l\_tot}(L, C) &= P_{l\_dr} + P_{l\_s} + P_{l\_on}(L) \\ &+ P_{l\_Cu}(L) + P_{l\_Fe}(L) + P_{l\_C}(L, C) \end{aligned} \tag{3.7}$$

The overall power loss must be as little as feasible since minimizing total power loss is comparable to increasing efficiency. As a result, **the first design goal is to reduce overall power loss while still maximizing power efficiency**.

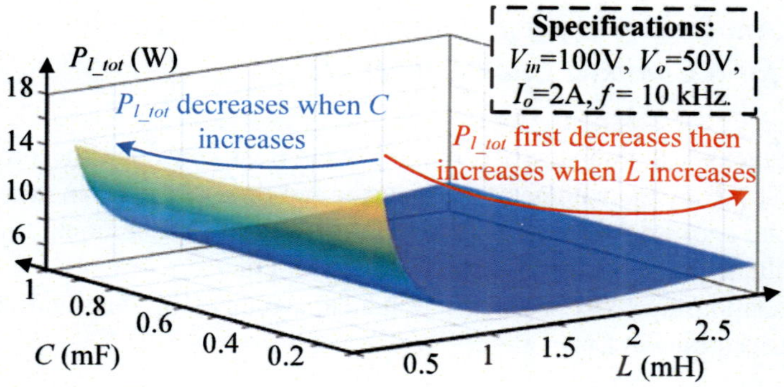

Fig. 3.7 Effects of L and C on total power loss P_{l_tot}

3.3.2 Analysis of Design Objective 2: Optimized Cut-Off Frequency for the Buck Converter with Optimal Filtering Capability

A decreased cut-off frequency indicates greater filtering effectiveness for the output *LC* filter in the Buck converter [18]. The relationships between f_c and L, and C are shown in Fig. 3.8 and (3.8). **The design objective 2 is to minimize the cut-off frequency for the Buck converter with optimized filtering capability**.

$$f_c(L, C) = \frac{1}{2\pi\sqrt{LC}} \tag{3.8}$$

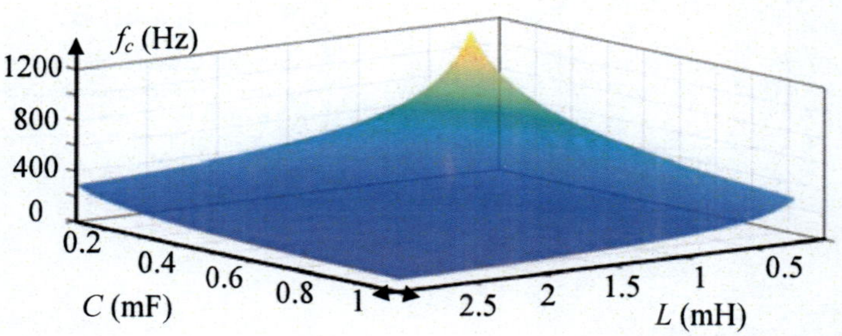

Fig. 3.8 Effects of L and C on cut-off frequency f_c

3.3.3 Analysis of Design Objective 3: Optimal Volume for a Compact Buck Converter

In applications with limited space, the size of the Buck converter is a crucial consideration [28]. Only the volume of the inductor and capacitor is taken into account in this chapter because the design parameters are L and C, and additional volumes such as the volume of the cooling system are taken for granted.

The relationship (3.9) between the inductor volume Vol_L and its inductance is deduced according to Liu et al. [18]:

$$Vol_L(L) = a_l \cdot L \tag{3.9}$$

where Vol_L is the inductor volume. a_l is computed by the linear regression method. For instance, inductors of the MCAP series of Multicomp with TAF-200 cores [29] are selected and shown in Fig. 3.9, in which statistical R^2 is close to 1, validating the linear relationship between Vol_L and L.

According to Liu et al. [18], the volume of capacitor Vol_C is linearly proportional to capacitance C, as (3.10) shows:

$$Vol_C(C) = a_c \cdot C \tag{3.10}$$

where Vol_C is the capacitor volume. a_c can be computed by the linear regression method. As an example, capacitors of the ECA1JM series of Panasonic [30] are selected in Fig. 3.10, in which statistical R^2 is close to 1, validating the linear relationship between Vol_C and C.

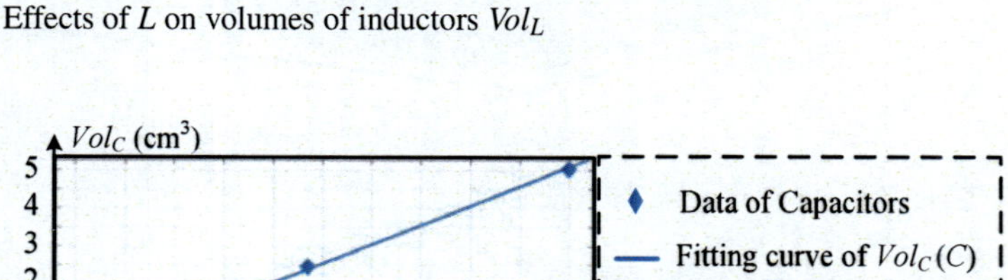

Fig. 3.9 Effects of L on volumes of inductors Vol_L

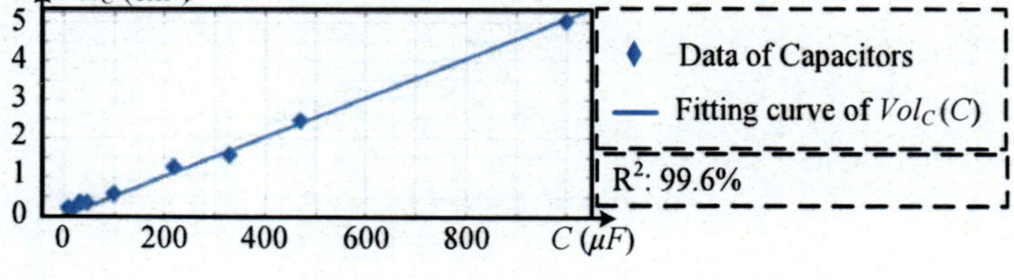

Fig. 3.10 Effects of C on volumes of capacitors Vol_C

With (3.9) and (3.10), the total volume V_{tot} to be minimized is shown in (3.11) and is linearly related to the inductance L and capacitance C. **The design objective 3 is to minimize the volume V_{tot} in (3.11) for a compact Buck converter.**

$$V_{tot}(L, C) = Vol_L(L) + Vol_C(C) = a_l \cdot L + a_c \cdot C \tag{3.11}$$

In summary, according to Figs. 3.7, 3.8, 3.9 and 3.10, as L and C increase, P_{l_tot} generally decreases f_c decreases, and V_{tot} increases. Thus, the minimization of volume V_{tot} is conflicting with the minimization of total power loss P_{l_tot} and cut-off frequency f_c.

3.4 The Proposed Multi-objective Design Approach for the Output *LC* Filter in Buck Converter with Coevolving AMOSA Algorithm

This section proposes a three-stage multi-objective design of output LC filter for Buck converter using coevolving-AMOSA algorithm to address problems I and II as stated in Sects. 3.2.2 and 3.2.3. Figure 3.11 describes the recommended design approach's flowchart.

3.4.1 Stage 1: Analysis of Three Design Objectives

As stated in the first section of Fig. 3.11, three conflicting objectives (P_{l_tot}, f_c, V_{tot}) are in-depth studied concerning L and C in Stage 1 of the proposed multi-objective design of the output LC filter for the Buck converter.

Based on the design conditions, P_{l_tot} can be analyzed with (3.1)–(3.7) in Sect. 3.3.1. f_c can be obtained with (3.8) in Sect. 3.3.2. And V_{tot} can be evaluated with (3.11) in Sect. 3.3.3. At the end of Stage 1, three objective functions regarding P_{l_tot}, f_c, and V_{tot} have been prepared for the multi-objective optimization in Stage 2.

3.4.2 Stage 2: Multi-objective Optimization of the Three Design Objectives with Coevolving-AMOSA Algorithm

3.4.2.1 Realization of Stage 2

The three design objectives will be multi-objective optimized using coevolving-AMOSA in Stage 2 using the three objective functions that were examined in Stage

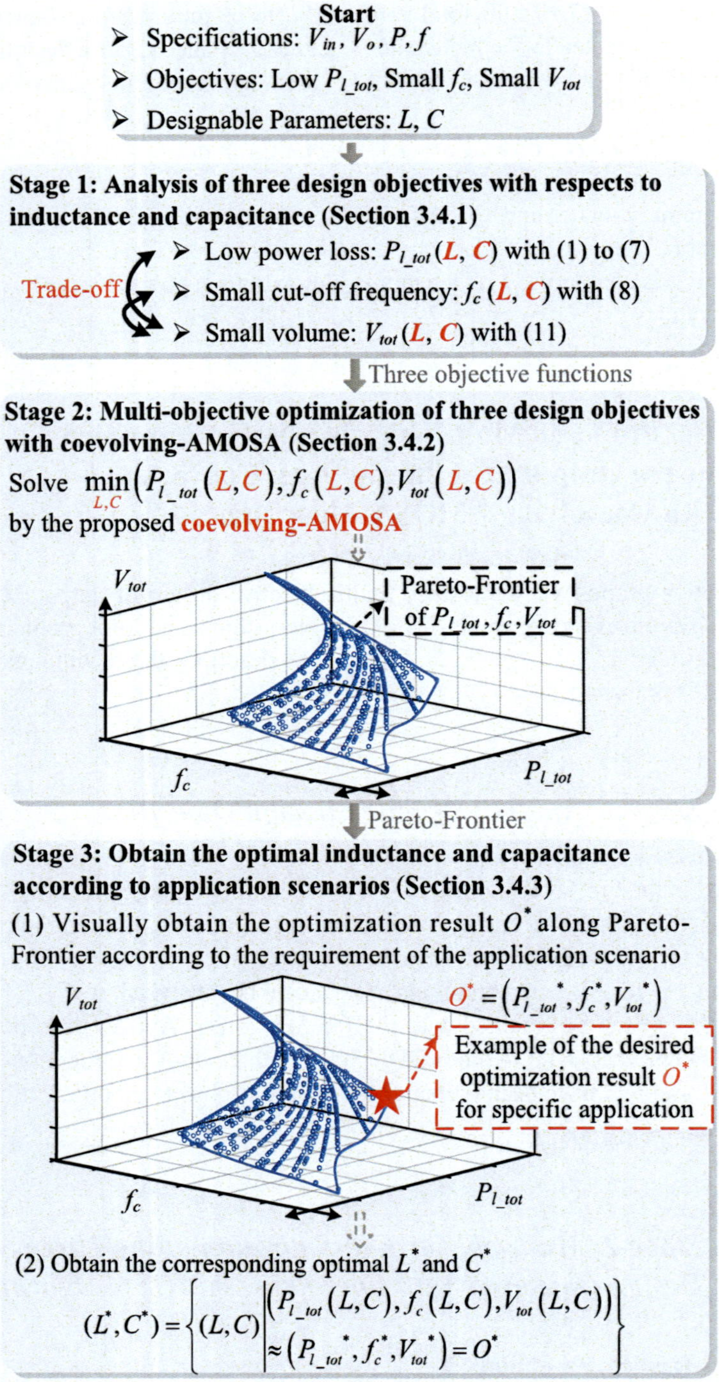

Fig. 3.11 The proposed three-stage multi-objective design of output *LC* filter for Buck converter with the coevolving-AMOSA algorithm

1 as shown in the second section of Fig. 3.11. The limitations of efficiency, cut-off frequency, and size are specified as $P_{l\_tot,\max}$, $f_{c,\max}$, and $V_{tot,\max}$, respectively, in (3.12), which also provides the optimization function for this issue.

$$\begin{aligned}
&\min_{L,C}(P_{l_tot}(L, C), f_c(L, C), V_{tot}(L, C)) \\
&s.t. \; P_{l_tot}(L, C) \le P_{l_tot,\max} \\
&\quad f_c(L, C) \le f_{c,\max} \\
&\quad V_{tot}(L, C) \le V_{tot,\max}
\end{aligned} \tag{3.12}$$

An enhanced AMOSA method known as the coevolving-AMOSA is used to solve (3.12), and it is explained in Sect. 3.4.2.2. The AMOSA method is adopted and changed primarily because it performs computations more quickly than other multi-objective algorithms such as NSGA-II [19], MOPSO [20], IBEA [22], etc. The proposed coevolving-AMOSA method makes it possible to reach a consistently and entirely covered Pareto-Frontier. The resulting Pareto-Frontier will then be given to Stage 3 for additional consideration when choosing the best design scenarios.

3.4.2.2 The Proposed Coevolving-AMOSA Algorithm

1. Flowchart of the proposed coevolving-AMOSA algorithm

   The flowchart of the proposed coevolving-AMOSA is compared with the traditional AMOSA [21] and shown in Fig. 3.12. The improvements of coevolving-AMOSA have been highlighted in red (steps 2 and 5). Table 3.1 also includes the pseudo-code for the coevolving-AMOSA that has been proposed, where U(0, 1) denotes the uniform distribution between [0, 1].
2. The advantages of the coevolving-AMOSA: improve uniformity and completeness of the obtained Pareto-Frontier

   The issues of Figs. 3.4 and 3.5 can be lessened since the coevolving-AMOSA method obtains the Pareto-Frontier, which has greater uniformity and completeness than the classic AMOSA.

The proposed coevolving-AMOSA method has two advantages: improved homogeneity of the Pareto-Frontier obtained between step 2 and step 5. A coevolving probability pr_0 is introduced in step 2 to regulate the creation of new designs (*L*, *C*). As the iterations go, pr_0 falls from 1 to 0. The produced new (*L*, *C*) will be directed toward the sparse areas within the Pareto-Frontier if a random number is greater than pr0. Step 5 also involves randomly removing excess designs (*L*, *C*) from congested places. The uniformity of the obtained Pareto-Frontier can be significantly increased with these two processes.

The second benefit, greater Pareto-Frontier completeness, can be attributed to step 2 of the coevolving-AMOSA algorithm. If a random value in step 2 is less than pr0, a new (*L*, *C*) will be created at random. This encourages the algorithm to extensively

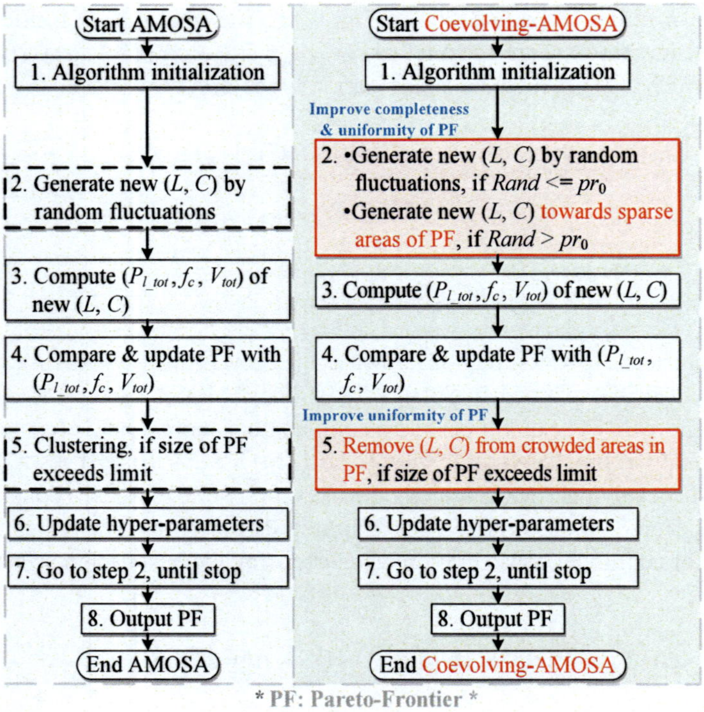

Fig. 3.12 Flowcharts of the traditional AMOSA algorithm [21] and the proposed coevolving-AMOSA algorithm (in which PF represents Pareto-frontier)

Table 3.1 Pseudo-code of the coevolving-AMOSA algorithm

Algorithm: the coevolving-AMOSA	
1	Initialize parameters of traditional AMOSA, coevolving probability pr_0, and PF to store optimal designs;
2	*Rand* ← U(0, 1);
3	IF $pr_0 > = Rand$
4	Generate new (L, C) with random fluctuations;
5	ELSE
6	Generate new (L, C) towards sparse areas in PF;
7	Compute (P_{l_tot}, f_c, V_{tot}) with (1)–(11);
8	IF size of Pareto-Frontier exceeds limit
9	Randomly remove extra (L, C) from crowded areas in PF;
10	Update parameters of traditional AMOSA, decrease pr_0;
11	IF stop criterion is not met
12	Go to line 2
13	OUTPUT PF

search the solution space to increase coverage, which improves the completeness of the Pareto-Frontier that is ultimately produced.

Therefore, the multi-objective design for the output *LC* filter in the Buck converter will be more precise and totally optimized with the uniform and complete Pareto-Frontier accomplished by the coevolving-AMOSA algorithm.

3. Comparisons between the proposed coevolving-AMOSA and other popular multi-objective algorithms

   In this part, several popular multi-objective evolutionary algorithms such as NSGA-II [19], MOPSO [20], traditional AMOSA [21], and some state-of-the-art algorithms such as IBEA [22] are given for comparison. These algorithms are run 30 times and are aimed at the multi-objective design in (3.12) for the output *LC* filter in the Buck converter. To indicate the uniformity and completeness of the Pareto-Frontier, two metrics are usually adopted: minimal spacing (SP_m), and maximum distribution range (*MDR*) [23]. Lower SP_m means better uniformity, and higher *MDR* means more complete coverage. Figure 3.13 shows a list of the comparison results.

As shown in Fig. 3.13a, the traditional AMOSA is the worst in terms of uniformity due to its clustering step 5 in Fig. 3.12 [21]. The Pareto-Frontier produced by the proposed coevolving-AMOSA is even more uniform than the Pareto-Frontier produced by the most advanced IBEA, as shown by the fact that SP_m shrinks with the proposed coevolving-AMOSA.

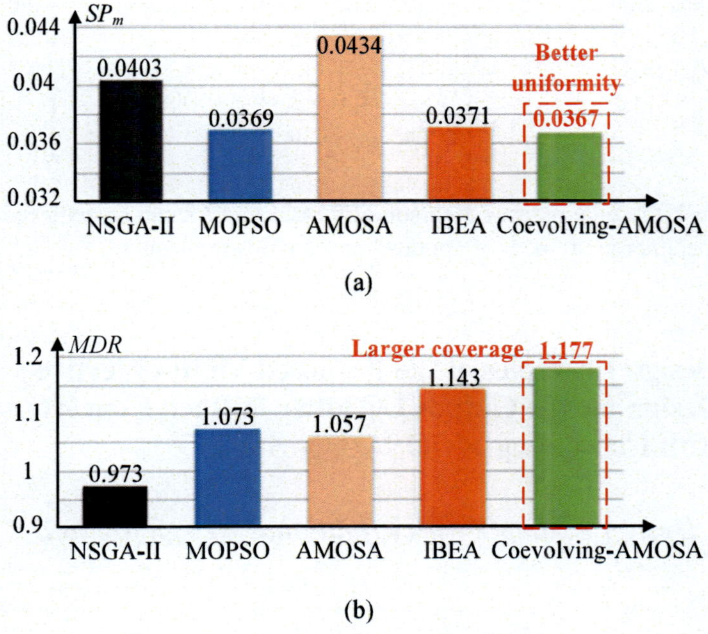

Fig. 3.13 Expected performance of multi-objective algorithms: **a** SP_m; **b** *MDR*

As shown in Fig. 3.13b, the *MDR* of NSGA-II is the lowest, meaning its coverage is far from satisfactory [31]. The coevolving-AMOSA shows its *MDR* at 1.177, resulting in a Pareto-Frontier that spans a larger region and offers engineers completely optimum design options. According to Fig. 3.13b, the coverage of the proposed coevolving-AMOSA is much bigger and greater than the cutting-edge algorithm IBEA.

As a result, the proposed coevolving-AMOSA method can provide a Pareto-Frontier with greater uniformity and completeness, resulting in designs for the output *LC* filter in Buck converters that are more precise and completely optimized.

3.4.3 Stage 3: Obtain the Optimal Design Solution Based on Application Requirements

Stage 3 of the proposed design strategy aims to find the ideal L and C for certain application situations using the Pareto-Frontier created in Stage 2. Stage 3 consists of the following 2 phases, as indicated in Fig. 3.11.

The Pareto-Frontier of three design objectives (power loss, cut-off frequency, and volume) is visually acquired at the beginning of Stage 3 to arrive at the optimization result $O^* = (P_{l\_tot}{}^*, f_c{}^*, V_{tot}{}^*)$ in accordance with the application scenario's requirements.

After that, with the picked optimization result O^*, the corresponding optimal L^* and C^* are obtained by (3.13) to find the combination of L and C which can meet O^* best.

$$(L^*, C^*) = \left\{ (L, C) \middle| \begin{array}{l} \left(P_{l\_tot}(L, C), f_c(L, C), V_{tot}(L, C)\right) \\ \approx \left(P^*_{l\_tot}, f^*_c, V^*_{tot}\right) = O^* \end{array} \right\} \tag{3.13}$$

Overall, the final optimization solution of L^* and C^* can be achieved according to the specific requirement of application scenarios in Stage 3.

3.5 Design Examples of the Proposed Multi-objective Design for the Output LC Filter in Buck Converter with Coevolving-AMOSA Algorithm

3.5.1 Design Example with Traditional Design Method

$$L = \frac{(1 - D)V_o}{f \cdot I_o \cdot I_{ripple}} \tag{3.14}$$

$$C = \frac{(\pi + 4 \cdot \tan \delta) \cdot I_{ripple} \cdot I_o}{8\pi f \cdot V_o \cdot V_{ripple}} \tag{3.15}$$

The conventional design is provided here for comparison with the proposed multi-objective design of the output *LC* filter for the Buck converter [8]. I_{ripple} and V_{ripple} are set as 40%, and 10% respectively.

With (3.14), the inductance is computed as 3.13 mH [8]. According to (3.15), the capacitance of the traditional design is computed as 30 μF [8]. P_{l_tot}, f_c, and V_{tot} of the traditional design are evaluated as 6.5 W, 495 Hz, and 122 cm^3.

3.5.2 Design Examples with the Proposed Multi-objective Design of Output LC Filter for Buck Converter with Coevolving-AMOSA Algorithm

3.5.2.1 Stage 1 of the Design Examples

In Table 3.2, the design specifications are presented. With respect to *L* and *C*, three objectives can be examined using (3.1)–(3.11), and they can be summed up as follows:

- *Objective*-1: minimize P_{l_tot} for the design of a high-efficiency Buck converter based on (3.1)–(3.7);
- *Objective*-2: minimize f_c for the design of a Buck converter with optimal filtering capability based on (3.8);
- *Objective*-3: minimize V_{tot} for the design of a compact Buck converter based on (3.9)–(3.11).

Table 3.2 Design specifications of design examples

Parameters of the main circuit			
P_o	100 W	f	10 kHz
V_{in}	100 V	V_o	50 V
Main switches (C2M0080120D)			
Q_g	62 nC	V_{gs}	20 V
R_{on}	80 mΩ	V_{SD}	4.3 V
Output LC filter			
Inductor cores	TAF-200 series		
Capacitor	100 V ECA1JM series of Panasonic		

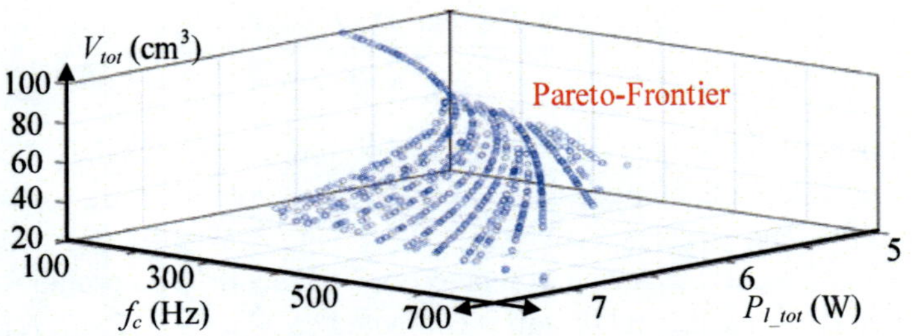

Fig. 3.14 Pareto-frontier (blue points) of power loss, cut-off frequency, and volume

3.5.2.2 Stage 2 of the Design Examples

The multi-objective design issue can be described in (3.12), where the $P_{l\_tot,\max}$ is set at 10 W, the $f_{c,\max}$ is set at 700 Hz, and the $V_{tot,\max}$ is set at 100 cm^3. This is based on the three design objective functions that were examined in Stage 1.

The Pareto-Frontier of power loss, cut-off frequency, and volume is constructed by following the pseudo-code of the proposed coevolving-AMOSA algorithm in Table 3.1, as illustrated in Fig. 3.14.

3.5.2.3 Stage 3 of the Design Examples

The three distinct application situations in Fig. 3.1 that are used as a basis for Stage 3's design cases are as follows.

Case 1: Reduce overall power loss while designing a high-efficiency Buck converter. Case 1 is appropriate for applications like satellites, ferries, airplanes, etc.

Case 2: In order to create a small Buck converter, the efficiency of the classic design must be maintained while the volume is reduced. Case 2 is suitable for small portable equipment including battery adapters, rooftop solar panels, digital cameras, LEDs, etc.

Case 3: Reduce cut-off frequency to create a Buck converter with the best filtering performance. Case 3 is suited to regions with stronger ripple reduction requirements, such as power audio amplifiers, MP3 players, and audio systems.

Optimization results O^* of the three design cases, which consist of $P_{l\_tot}{}^*$, $f_c{}^*$ and $V_{tot}{}^*$, are visually obtained from the Pareto-Frontier in Fig. 3.14 and listed in Table 3.3.

The 3-D Pareto-Frontier in Fig. 3.14 is projected into three 2-D plots, as shown in Fig. 3.15, along with the conventional design and three ideal design examples, for better visibility. According to Fig. 3.15, case 1 reduces power loss by 1.32 W compared to the conventional design, case 2 maintains the conventional design's efficiency while reducing volume by 92.9 cm^3, and case 3 reduces the cut-off frequency by 392 Hz.

Table 3.3 Objective values of the three required optimal designs

	$P_{l\_tot}^*$ (W)	f_c^* (Hz)	V_{tot}^* (cm^3)
Optimal case 1	5.18	146	63.4
Optimal case 2	6.46	323	29.1
Optimal case 3	5.7	103	100.0

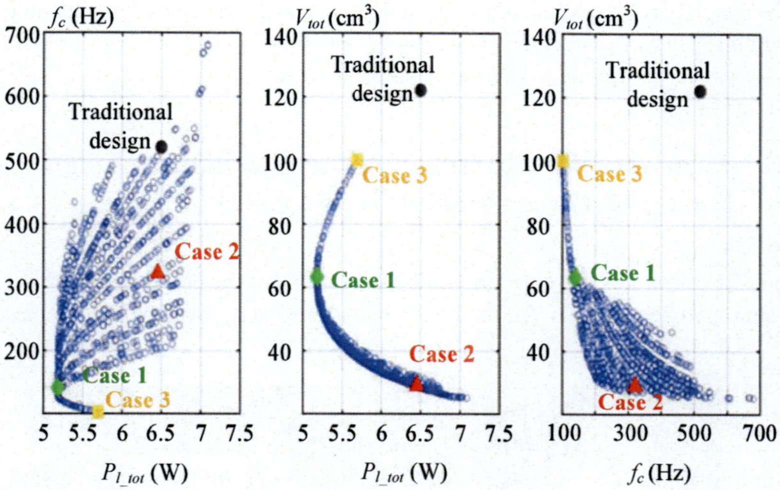

Fig. 3.15 Projected Pareto-frontier (blue points): **a** power loss versus cut-off frequency; **b** power loss versus volume; **c** cut-off frequency versus volume

After that, with (3.13) and the optimization results O^* of the three design cases in Table 3.3, the corresponding optimization solutions L^* and C^* are obtained and listed in Table 3.4.

Table 3.4 Inductance and capacitance of three optimal designs

	Inductance L^* (mH)	Capacitance C^* (μF)
Optimal design case 1	1.49	800
Optimal design case 2	0.688	352
Optimal design case 3	2.41	1000

3.6 Experimental Verification

Hardware experiments in this part are used to test the viability and efficacy of the proposed multi-objective design strategy for the output *LC* filter in the Buck converter with the coevolving-AMOSA algorithm. The design examples are provided in Table 3.4 in Sect. 3.5. In Fig. 3.16, the hardware main circuit is displayed. Table 3.5 displays the thorough hardware realization of Table 3.4. The design requirements match those in Table 3.2.

3.6.1 Experimental Waveforms of the Traditional Design Case and Three Optimal Design Cases

Figure 3.17 provides the experimental waveforms for the conventional design situation shown in Sect. 3.5.1a. And in Fig. 3.17b–d, respectively, are the three best design

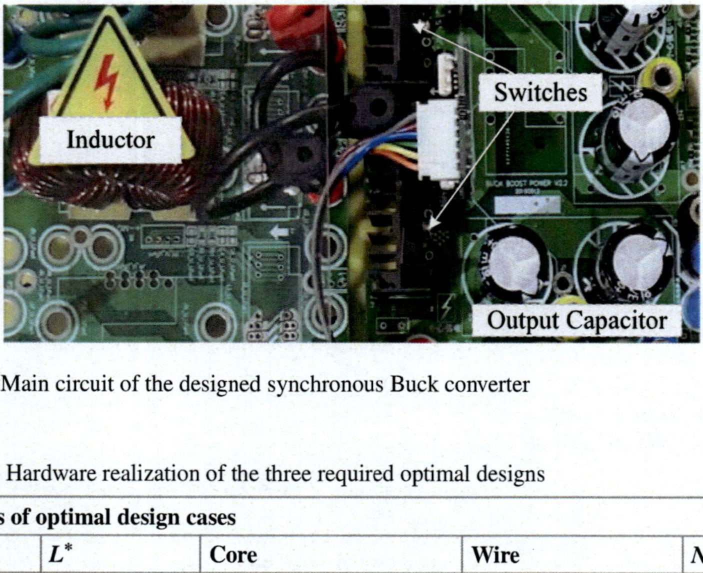

Fig. 3.16 Main circuit of the designed synchronous Buck converter

Table 3.5 Hardware realization of the three required optimal designs

Inductors of optimal design cases				
Cases	***L******	**Core**	**Wire**	***N***
Case 1	1.49 mH	T200B-75-200	UEFN/U 1 mm	97
Case 2	688 μH	T175-75-200	UEFN/U 1 mm	81
Case 3	2.41 mH	T250-75-200	UEFN/U 1 mm	100

Capacitors of optimal design cases		
Cases	***C**** (μF)	**Realization**
Case 1	800	Parallel: 330 and 470 μF, 100 V ECA1JM
Case 2	352	Parallel: 22 and 330 μF, 100 V ECA1JM
Case 3	1000	Single: 1000 μF, 100 V ECA1JM

instances using the proposed multi-objective design strategy of output LC filter in Buck Converter via coevolving-AMOSA algorithm shown in Sect. 3.5.2.

3.6.2 *Evaluation of the Experimental Results*

According to total power loss, cut-off frequency, and volume, the performance indicators of the conventional and three optimal design instances are analyzed in experiments and listed in Fig. 3.18. The following is a list of the extensive evaluations.

3.6.2.1 Traditional Design Case

The standard design method described in Sect. 3.5.1 results in a 123 cm^3 volume for the designed output *LC* filter in the Buck converter. The efficiency is 93.05% and the total power loss is 6.8 W since the input and output powers are 97.9 W and 91.1 W, respectively. 490 Hz is its cut-off frequency.

3.6.2.2 Optimal Design Case 1: Maximizing Efficiency

It is anticipated that the best design case 1, as introduced in Sect. 3.5.2.3, will have the least amount of power loss. According to the testing findings, the best design case 1 has a capacity that is 51.5% less than the standard design at 59.6 cm^3. Its input and output powers are 98.1 W and 92.8 W, respectively. As a result, its efficiency is 94.6% and its overall power loss is 5.3 W. Comparing the optimal design case 1 to the conventional design case results in a 1.6 W loss savings. Additionally, it has a 147 Hz cutoff frequency. As a result, the high-efficiency applications in Fig. 3.1 are appropriate for the ideal design case 1 such as electric cars, airplanes, etc.

3.6.2.3 Optimal Design Case 2: Minimizing Volume While Maintaining Same Efficiency as the Traditional Design

It is anticipated that the optimal design case 2, as described in Sect. 3.5.2.3, will have a reduced volume without sacrificing its power efficiency. According to the experimental findings, the ideal design case 2 has input and output powers of 96.4 and 89.7 W, respectively. As a result, total power loss is 6.7 W and efficiency is 93.0%, which is nearly identical to the conventional design case. Its volume is 29 cm^3 less than the conventional design, a reduction of 76.4%. 323 Hz is the cut-off frequency. Therefore, space-constrained portable equipment like battery adapters, rooftop PV, digital cameras, LED, etc. can use the best design case 2.

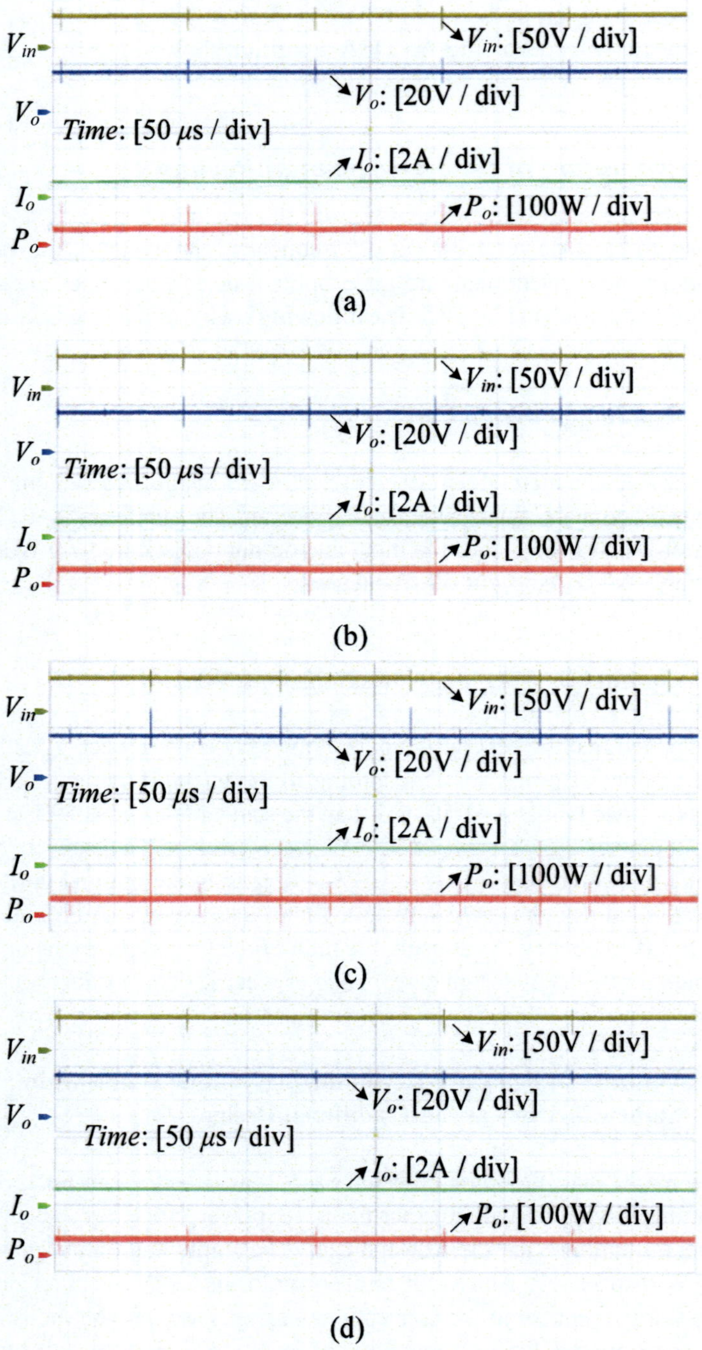

Fig. 3.17 Waveforms of the design cases: **a** traditional design; **b** optimal design case 1; **c** optimal design case 2; **d** optimal design case 3

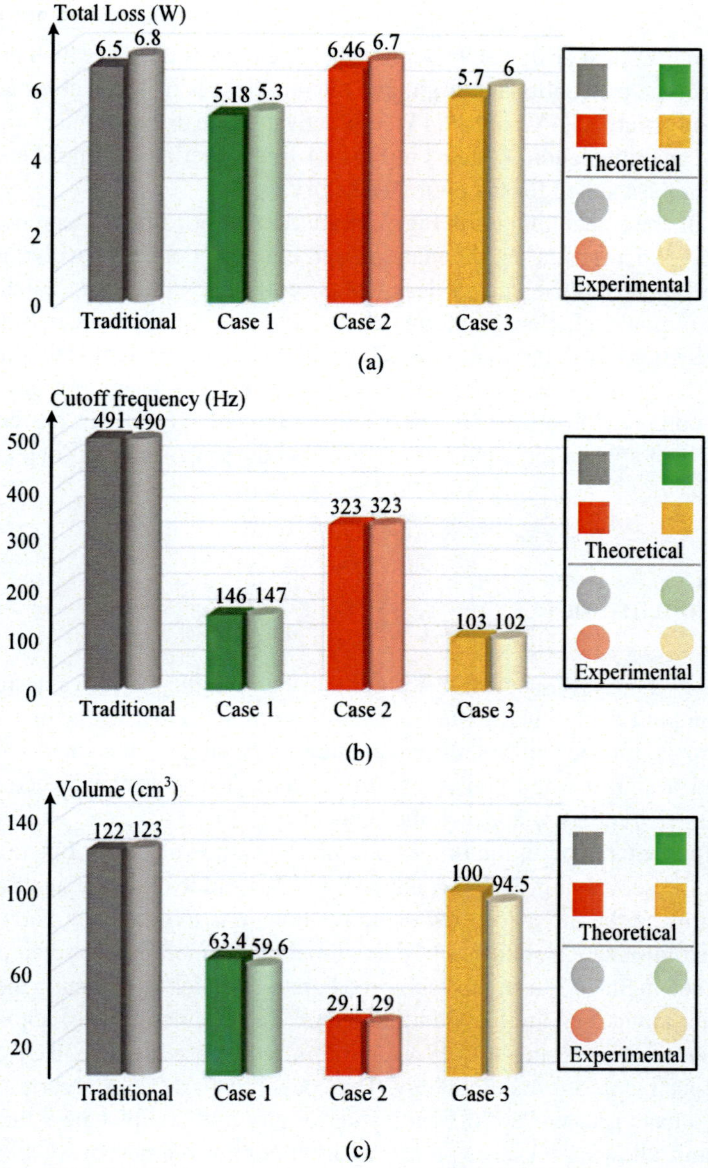

Fig. 3.18 Experimental and theoretical results of the conventional and optimal design cases: **a** total power loss; **b** cut-off frequency; **c** volume

3.6.2.4 Optimal Design Case 3: Minimizing Cut-Off Frequency

For the best filtering performance, the optimal design scenario 3, as stated in Sect. 3.5.2.3, is anticipated to have a minimized cut-off frequency. According to the testing findings, the best design case 3 has a cut-off frequency that is 79.2%

lower than the conventional design at 102 Hz. Its volume is also 23.2% lower than that of conventional design at 94.5 cm^3. The efficiency is 93.7%, which is somewhat better than the conventional design, and the total loss is 6 W because the input and output powers are 95.4 W and 89.4 W, respectively. Due to the tougher constraints on filtering capability, audio systems in Fig. 3.1 like power audio amplifiers and MP3 players are best suited for the optimal design case 3.

Overall, the actual findings in Fig. 3.18 are consistent with the theoretical analysis in Table 3.3, demonstrating the viability and efficiency of the coevolving-AMOSA algorithm-based multi-objective design of the output *LC* filter for the Buck converter. In terms of power efficiency, filtering ability, and volume, the three optimum design examples outperform the conventional design example, demonstrating the optimal designs' fully-optimized performance with the proposed design process. Additionally, the output *LC* filters can be adaptably constructed to fulfill diverse requirements in a variety of application scenarios according to the proposed multi-objective design approach.

3.7 Conclusion

In Chap. 5, a coevolving-AMOSA algorithm-based multi-objective design strategy for the output *LC* filter in the Buck converter is proposed to address three competing design goals: low power loss, improved filtering capability, and small volume. This proposed design method consists of three steps. Three design objectives (power loss, cut-off frequency, and volume) concerning inductance and capacitance will be carefully examined in the first step to produce three objective functions. And in Stage 2, the coevolving-AMOSA method will be used to optimize a multi-objective problem using the three acquired objective functions to produce a Pareto-Frontier. The final optimization solutions of the optimal inductance and capacitance will then be obtained in Stage 3, using the achieved Pareto-Frontier, by selecting the optimization result along the Pareto-Frontier based on the specific needs of applications. In particular, the coevolving-AMOSA algorithm is proposed for use in Stage 2 of this multi-objective design methodology. The coevolving-AMOSA algorithm has been shown to have greater Pareto-Frontier uniformity and completeness than previous algorithms, allowing for more precise and fully optimized design solutions.

The proposed multi-objective design technique for the output *LC* filter in the Buck converter via the coevolving-AMOSA algorithm has been shown with three optimal design examples based on various needs in three application situations. Through hardware tests and comparisons with the design example created using the conventional design process, the optimized performance of these three optimal design examples has been confirmed. The proposed multi-objective design strategy for the output *LC* filter in the Buck converter with the coevolving-AMOSA algorithm has thus been proven to be feasible and successful.

References

1. S.-K. Lim, H.-S. Lee, H.-R. Cha, and S.-J. Park, 'Multi-Level DC/DC Converter for E-Mobility Charging Stations', *IEEE Access*, vol. 8, pp. 48774–48783, 2020.
2. H. U. R. Habib, S. Wang, M. R. Elkadeem, and M. F. Elmorshedy, 'Design Optimization and Model Predictive Control of a Standalone Hybrid Renewable Energy System: A Case Study on a Small Residential Load in Pakistan', *IEEE Access*, vol. 7, pp. 117369–117390, 2019.
3. G. Zhang, N. Jin, L. Qu, and S. S. Yu, 'Inherently Non-Pulsating Input Current DC-DC Converter for Battery Storage Systems', *IEEE Access*, pp. 1–1, 2020.
4. H. You and X. Cai, 'A Three-Level Modular DC/DC Converter Applied in High Voltage DC Grid', *IEEE Access*, vol. 6, pp. 25448–25462, 2018.
5. J. M. Martinez-Heredia, F. Colodro, J. L. Mora-Jimenez, A. Remujo, J. Soriano, and S. Esteban, 'Development of GaN Technology-Based DC/DC Converter for Hybrid UAV', *IEEE Access*, vol. 8, pp. 88014–88025, 2020.
6. M. Sobaszek, 'Self-Tuned Class-D Audio Amplifier With Post-Filter Digital Feedback Implemented on Digital Signal Controller', *IEEE Trans. Circuits Syst. I*, vol. 67, no. 3, pp. 797–805, Mar. 2020.
7. B. Chandrasekar *et al.*, 'Non-Isolated High-Gain Triple Port DC–DC Buck-Boost Converter With Positive Output Voltage for Photovoltaic Applications', *IEEE Access*, vol. 8, pp. 113649–113666, 2020.
8. R. W. Erickson, *Fundamentals of power electronics*. New York: Chapman & Hall, 1997.
9. A. De Nardo, N. Femia, G. Petrone, and G. Spagnuolo, 'Optimal Buck Converter Output Filter Design for Point-of-Load Applications', *IEEE Transactions on Industrial Electronics*, vol. 57, no. 4, pp. 1330–1341, Apr. 2010.
10. J. Jiang, Z. Li, K. Song, B. Song, S. Dong, and C. Zhu, 'A Cascaded Topology and Control Method for Two-Phase Receivers of Dynamic Wireless Power Transfer Systems', *IEEE Access*, vol. 8, pp. 47445–47455, 2020.
11. L. Hu, X. Wei, Y. Luo, and J. Ma, 'Synchronous High Step-Down Ratio Non-Isolated LED Constant Current Driver Based on Improved Watkins-Johnson Topology', *IEEE Access*, vol. 7, pp. 18345–18353, 2019.
12. R. Burkart and J. W. Kolar, 'Component cost models for multi-objective optimizations of switched-mode power converters', in *2013 IEEE Energy Conversion Congress and Exposition*, Denver, CO, USA, Sep. 2013, pp. 2139–2146.
13. P. Pelletier, J.-M. Guichon, J.-L. Schanen, and D. Frey, 'Optimization of a DC Capacitor Tank', *IEEE Transactions on Industry Applications*, vol. 45, no. 2, pp. 880–886, 2009.
14. D. O. Boillat, F. Krismer, and J. W. Kolar, 'Design Space Analysis and ρ–η Pareto Optimization of LC Output Filters for Switch-Mode AC Power Sources', *IEEE Transactions on Power Electronics*, vol. 30, no. 12, pp. 6906–6923, Dec. 2015.
15. H. Wang *et al.*, 'Transitioning to Physics-of-Failure as a Reliability Driver in Power Electronics', *IEEE J. Emerg. Sel. Topics Power Electron.*, vol. 2, no. 1, pp. 97–114, Mar. 2014.
16. X. Li, F. Lin, X. Zhang, M. Huang, and H. Wang, 'Multi-objective Design of LC Filter for High-efficiency, High-power-density and High-performance Buck Converter', in *2019 IEEE Energy Conversion Congress and Exposition (ECCE)*, Baltimore, MD, USA, Sep. 2019, pp. 5132–5136.
17. Hyosung Kim and Seung-Ki Sul, 'Analysis on output LC filters for PWM inverters', in *2009 IEEE 6th International Power Electronics and Motion Control Conference*, Wuhan, su 2009, pp. 384–389.
18. Y. Liu, M. Huang, H. Wang, X. Zha, J. Gong, and J. Sun, 'Reliability-Oriented Optimization of the LC Filter in a Buck DC-DC Converter', *IEEE Transactions on Power Electronics*, vol. 32, no. 8, pp. 6323–6337, Aug. 2017.
19. Q. Bian, B. Nener, and X. Wang, 'A Modified NSGA-II for Solving Control Allocation Optimization Problem in Lateral Flight Control System for Large Aircraft', *IEEE Access*, vol. 7, pp. 17696–17704, 2019.

20. A. Sundaram, 'Combined Heat and Power Economic Emission Dispatch Using Hybrid NSGA II-MOPSO Algorithm Incorporating an Effective Constraint Handling Mechanism', *IEEE Access*, vol. 8, pp. 13748–13768, 2020.
21. S. Bandyopadhyay, S. Saha, U. Maulik, and K. Deb, 'A Simulated Annealing-Based Multiobjective Optimization Algorithm: AMOSA', *IEEE Transactions on Evolutionary Computation*, vol. 12, no. 3, pp. 269–283, Jun. 2008.
22. Q. Xu, Z. Xu, and T. Ma, 'A Survey of Multiobjective Evolutionary Algorithms Based on Decomposition: Variants, Challenges and Future Directions', *IEEE Access*, vol. 8, pp. 41588–41614, 2020.
23. S. Bandyopadhyay, S. K. Pal, and B. Aruna, 'Multiobjective GAs, Quantitative Indices, and Pattern Classification', *IEEE Transactions on Systems, Man and Cybernetics, Part B (Cybernetics)*, vol. 34, no. 5, pp. 2088–2099, Oct. 2004.
24. M. Zhang and Y. Li, 'Multi-Objective Optimal Reactive Power Dispatch of Power Systems by Combining Classification-Based Multi-Objective Evolutionary Algorithm and Integrated Decision Making', *IEEE Access*, vol. 8, pp. 38198–38209, 2020.
25. N. Kondrath and M. K. Kazimierczuk, 'Inductor winding loss owing to skin and proximity effects including harmonics in non-isolated pulse-width modulated dc–dc converters operating in continuous conduction mode', *IET Pwr. Electr.*, vol. 3, no. 6, p. 989, 2010.
26. J. Reinert, A. Brockmeyer, and R. W. A. A. De Doncker, 'Calculation of losses in ferro- and ferrimagnetic materials based on the modified Steinmetz equation', *IEEE Trans. on Ind. Applicat.*, vol. 37, no. 4, pp. 1055–1061, Aug. 2001.
27. M. R. Banaei and H. A. F. Bonab, 'A High Efficiency Nonisolated Buck–Boost Converter Based on ZETA Converter', *IEEE Trans. Ind. Electron.*, vol. 67, no. 3, pp. 1991–1998, Mar. 2020.
28. X. Lu and H. Wang, 'A Highly Efficient Multifunctional Power Electronic Interface for PEV Hybrid Energy Management Systems', *IEEE Access*, vol. 7, pp. 8964–8974, 2019.
29. Multicomp, "Toroidal Inductor, MCAP Series", MCAP115018077A-561LU datasheet, Feb. 2020.
30. Panasonic, "Aluminum Electrolytic Capacitors", ECA1JM331 datasheet, Dec. 2019.
31. K. Deb, A. Pratap, S. Agarwal, and T. Meyarivan, 'A fast and elitist multiobjective genetic algorithm: NSGA-II', *IEEE Trans. Evol. Computat.*, vol. 6, no. 2, pp. 182–197, Apr. 2002.

Chapter 4
The Proposed Artificial-Intelligence-Based Design (AI-D) for Circuit Parameters of Power Converters

4.1 Introduction

Power converters are currently used more frequently in both industry and day-to-day living. Power converters can control power transmission and change the voltage and current's shape [1]. In industrial applications, power converters such as DC-DC converters and inverters are the critical enablers in renewable energy systems [2], wireless power transfer [3], and DC microgrids [4]. Even in our daily life, the applications of power converters are omnipresent, such as electric vehicles [5], solar PV [6], etc.

Power converter circuit characteristics should be carefully specified in order to guarantee the good performance of power converters in all applications. And efficiency, size, cost, reliability, ripples, and transient response are several commonly-adopted design objectives [7–10]. The parameter design for power converters is significant and difficult since, in order to get fundamentally superior performance, typically numerous design objectives are taken into account simultaneously.

The process of analysis and deduction is followed by the process of optimization when designing parameters for power converters. Up to now, there have been two main approaches in the parameter design of power converters, which are the traditional human-dependent design approach [11, 12] and the computer-aided optimization (CAO) design approach [8, 13, 14]. For the conventional human-dependent design approach, engineers completely analyze optimization objectives and design constraints to derive mathematical expressions [11]. Additionally, the optimization procedure will involve numerous manual attempts and errors. The traditional human-dependent design technique has several limitations, including a high workload and low precision brought on by approximation during the analysis and deduction process [11]. Regarding the CAO approach, which was proposed in the last two decades, the optimization process was carried out using some optimization algorithms on computers, such as the recently popular particle swarm optimization algorithm,

X. Zhang et al., *Automated Design of Electrical Converters with Advanced AI Algorithms*,
https://doi.org/10.1007/978-981-99-0459-4_4

genetic algorithm, and ant colony algorithm [14]. However, there is still a considerable level of human dependence in the analysis and deduction process for optimization targets and design constraints, leading to low accuracy and a significant amount of time required [13].

Neural networks (NN), an artificial intelligence (AI) approach, are the best option for relieving the labor-intensive analytical and deduction stages of parameter design. The ability of NN to learn from and comprehend external data is achieved by tweaking its adjustable weights, which mimic the adaptive connections of neurons in the brain. Any complex non-linear function can be accurately learned thanks to the NN's easily scalable structure. Due to these advantages, NN has been widely adopted in motor optimization [15], modulation strategies [16], and control [17, 18]. For instance, [15] adopts an extreme learning machine to fit the results of finite element analysis to optimize the motor. Filho et al. [16] utilizes NN to realize optimal selective harmonic elimination. NN serves as the model predictive controller for modular multilevel converters in [18].

However, the majority of the NNs used in the existing literature on power converters are simple networks, and as a result, their generalization accuracy for unobserved data could yet be improved. The structure of NN needs to be changed to achieve greater prediction accuracy with unobserved data.

In this chapter, an artificial intelligence-based design (AI-D) approach is examined to address the aforementioned challenges that the current parameter design approaches for power converters are encountering. This approach was motivated by the powerful learning capabilities of NN. The two methods in parameter designs that can be automated in the proposed AI-D methodology are the analysis and deduction process and the optimization process. Data-driven models will execute the analysis and deduction process with the aid of simulation tools and a batch-normalization neural network (BN-NN). And the optimization process will be handled by the evolutionary algorithm (EA). The parameter design for power converters can obtain precisely optimal design results and a wide degree of freedom for engineers with BN-NN and the automatic fashion in these two operations.

The remainder of this chapter is set up as follows. Section 4.2 describes the issues with the parameter design methodologies for power converters that are currently in use as well as the proposed fixes. The proposed AI-based design approach's step-by-step workflow is described in Sect. 4.3. In Sect. 4.4, a design instance is illustrated, and in Sect. 4.5, experimental findings are reported. The chapter's conclusion is provided in Sect. 4.6.

4.2 Problem Descriptions for the Parameter Design Approaches for Power Converters and the Proposed Solutions

4.2.1 Problems in the Existing Circuit Parameter Design Approaches for Power Converters

After defining the design goals and operating conditions, there are two key steps that must be taken when designing the circuit parameters for power converters. The first step is the analysis and deduction procedure for the optimization objectives and design constraints, from which the mathematical expressions for these two concepts will be derived. The second is an optimization procedure to identify the circuit characteristics that will achieve the optimization objective's optimum performance without violating the design limitations.

Engineers do the analysis and deduction process as well as the optimization process in the conventional human-dependent design approach, as shown on the left side of Fig. 4.1. For the sake of analytical ease, various approximations will be used in order to generate the mathematical equations for the intended optimization objective and design constraints. For example, to derive the operating point of the two-switch forward converter in [11], the switches are assumed to be ideal. Besides, as introduced by Lin et al., in the deduction of the output impedance of dual active bridge converter, only 0th and 1st order terms in Fourier expansion are considered [19]. Even though sometimes approximations are removed for the sake of high accuracy, the computational burden and complexity are heavy for engineers [12, 20]. When it comes to the optimization process, repetitious manual trials and errors are also time-consuming and have no guarantee for optimization accuracy [21].

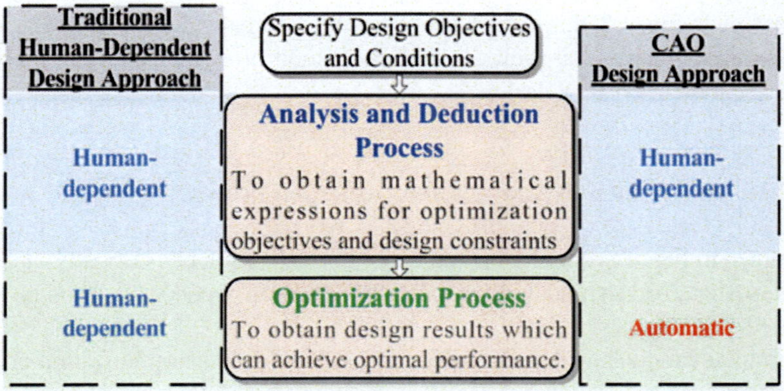

Fig. 4.1 Realization of design process (human-dependence or automation) in two kinds of parameter design approaches for power converters: traditional human-dependent design approach and CAO design approach

Over the previous 20 years, while computer science has advanced, the CAO parameter design approach—which is shown on the right side of Fig. 4.1—has undergone significant advancements. The CAO design approach has realized automation in the optimization process with the assistance of optimization algorithms [8, 13, 22–24]. For instance, [8] modifies the archived multi-objective simulated annealing algorithm (AMOSA) to achieve the multi-objective design of Buck converter. Neugebauer and Perreault [22] adopts Monte-Carlo for the optimal DC-DC converter in automobiles. In order to create a good comprehensive parameter design, Liu et al. examine the internal temperature rise of the capacitor to determine its lifetime, study the volume of L and C using the definitions of inductance and capacitance and their geometric features, and use the NSGA-II algorithm [25].

The human dependence in the analysis and deduction process is still a problem, despite the CAO design approach having liberated engineers from repetitive tries and errors to obtain optimal design solutions. To derive the mathematical expressions of the optimization objective and design constraints, it is necessary to conduct a difficult and time-consuming analysis that also raises questions about correctness. Furthermore, because CAO techniques use optimization algorithms that are essentially strict instructions to find the best possible solutions for a given function, they are unable to learn from and understand outside inputs in order to develop intelligence. As a result, CAO methods cannot be classified as AI in the strict sense.

4.2.2 The Proposed Solutions for the Automated Design for the Circuit Parameters of Power Converters

An artificial-intelligence-based design (AI-D) technique is specifically proposed to address the issues including excessive work burden and low precision in the parameter design for power converters, which are related to the high level of human dependence. According to Fig. 4.2, this design technique can automatically complete the analysis and deduction process as well as the optimization process, making it simpler and more accurate to create a design.

4.2.2.1 Automation in Analysis and Deduction Process

As shown in Fig. 4.3, simulation and batch-normalization NN (BN-NN) are used to develop a data-driven model in order to remove human dependence from the analysis and deduction process.

Simulated models will be created to assess how well the optimization objectives and design constraints work when various design parameters are chosen. A neural network will get a data set for training from simulation models. NN can be used as an accurate data-driven model after training with sparse performance data since it can deduce any complex and nonlinear correlations between design factors and

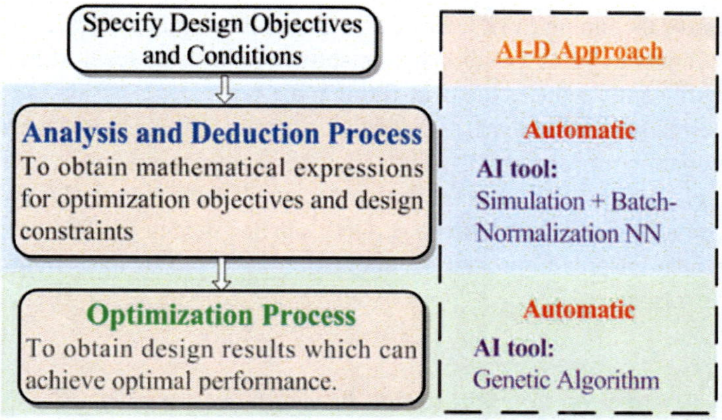

Fig. 4.2 Realization of design process in the proposed AI-D approach for the circuit parameters of power converters

Fig. 4.3 Automation in the analysis and deduction process in the proposed AI-D

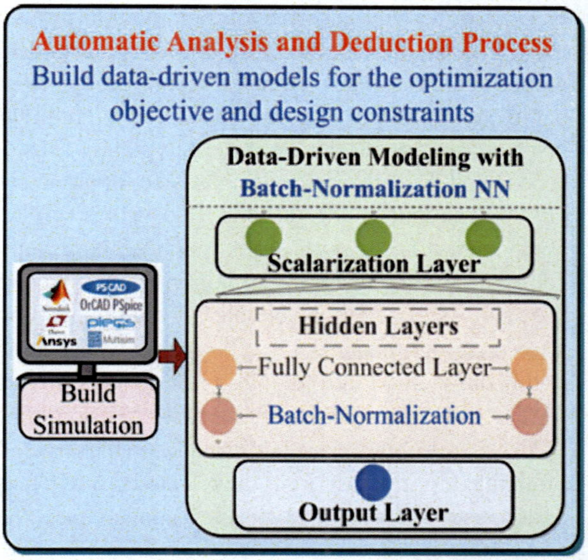

performance. As a result, it can be used to forecast how well new ideas would function.

For the purpose of creating data-driven models, BN-NN is specifically used to achieve improved generalization accuracy in forecasting unseen design performance. It has three different types of layers: the output layer, hidden layers, and scalarization layer. The particular batch-normalization layer in the buried layer gives BN-NN its name. The batch-normalization layer aims to avoid the over-fitting problem, which will seriously deteriorate the NN accuracy. Compared with other techniques for avoiding over-fitting, such as L2 weight-decay regularization [26] and dropout [27], the batch-normalization layer [28] does not introduce extra hyperparameters and thus is simpler for the tuning of NN. Basically, the batch-normalization layer normalizes the layer inputs and has the ability to adaptively adjust the level of normalization, which significantly reduces the over-fitting issue [28].

The design process analysis and deduction steps can be carried out automatically using simulation models and BN-NN, relieving engineers of a time-consuming task. Furthermore, compared to the conventional manual analysis procedure with approximation, a more accurate performance rating can be guaranteed. The unique BN-NN structure also ensures improved generalization accuracy in foretelling unforeseen design performance.

4.2.2.2 Automation in Optimization Process

The optimization phase, in which design results are finalized to achieve optimal performance, can also be automated in addition to the analysis and deduction processes.

One of the well-known evolutionary algorithms, the genetic algorithm (GA), is chosen to handle parameter optimization. This decision is taken into account the unique feature of the parameter design problem for power converters. The design parameters for a power converter can be in continuous space or discrete space. For instance, continuous design space is typically considered when switching frequency is designed. In contrast, the design space for inductors and capacitors is discrete since only discontinuous and disconnected values can be selected due to their practical components. The mix of design parameters in both discrete and continuous space contributes to a mixed-integer optimization problem. Among popular EA (PSO, GA, and ACO), PSO is suitable for continuous optimization [29], and ACO aims at discrete optimization [30]. Fortunately, GA performs the best in mixed-integer optimization and it also enjoys a fast convergence speed [23]. Thus, GA is suitable for the AI-D approach for the parameter design of power converters.

In order to obtain good performance in the targeted optimization objective without violating design restrictions, the selected GA will help identify global optimal design parameters. Additionally, its quick convergence speed aids in the quick design of power converters.

In summary, the proposed AI-D technique can achieve a high level of automation in the analysis and deduction process and the optimization process, as was previously discussed. High prediction accuracy in previously unidentified design parameters is ensured by the unique BN-NN used in AI-D. As a result, the proposed AI-D approach enables high precision and simple application in the power converter parameter design.

4.3 AI-D Approach for the Parameter Design of Power Converters

There are no restrictions on the types of power converters that can be designed with the proposed AI-D methodology in terms of application backgrounds. For the purpose of clarity, the proposed AI-D technique is validated in this chapter using the synchronous Buck converter's circuit parameter design in the 48–12 V accessory-load power supply system of an electric vehicle (EV) [31].

To achieve power conversion from the available 48 V power supply to the necessary 12 V accessory-load voltage in EV, the 48–12 V power supply system is being used more and more [32], as shown in Fig. 4.4. To facilitate this power conversion process, usually a synchronous Buck topology is adopted thanks to its good efficiency and simple structure [31, 32]. Efficiency is one crucial performance metric to consider when designing something to guarantee good power transfer efficiency. To ensure optimal overall performance, volume and ripples are also taken into account for the benefit of a compact and dependable design. The design process in this chapter uses efficiency as the optimization goal and uses volume and ripples as two examples of design constraints. It should be highlighted that the proposed AI-D approach can still be used and only minor adjustments are required if any other objectives or limitations, such as cost or dependability, are favored in various design contexts.

This section elaborates on the AI-D technique, which takes volume and ripple into account while designing the circuit parameters for the efficiency-oriented synchronous Buck converter in EV. According to Fig. 4.5, there are 4 steps in this design process.

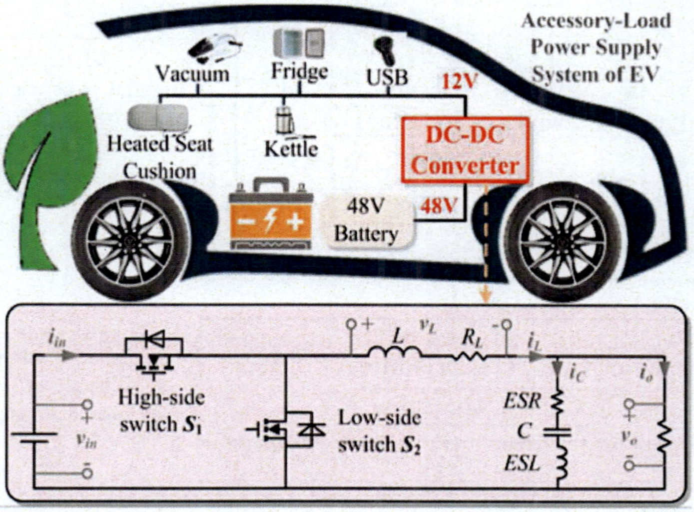

Fig. 4.4 The parameter design of synchronous Buck converter in the 48–12 V accessory-load power supply system in EV via the proposed AI-D approach

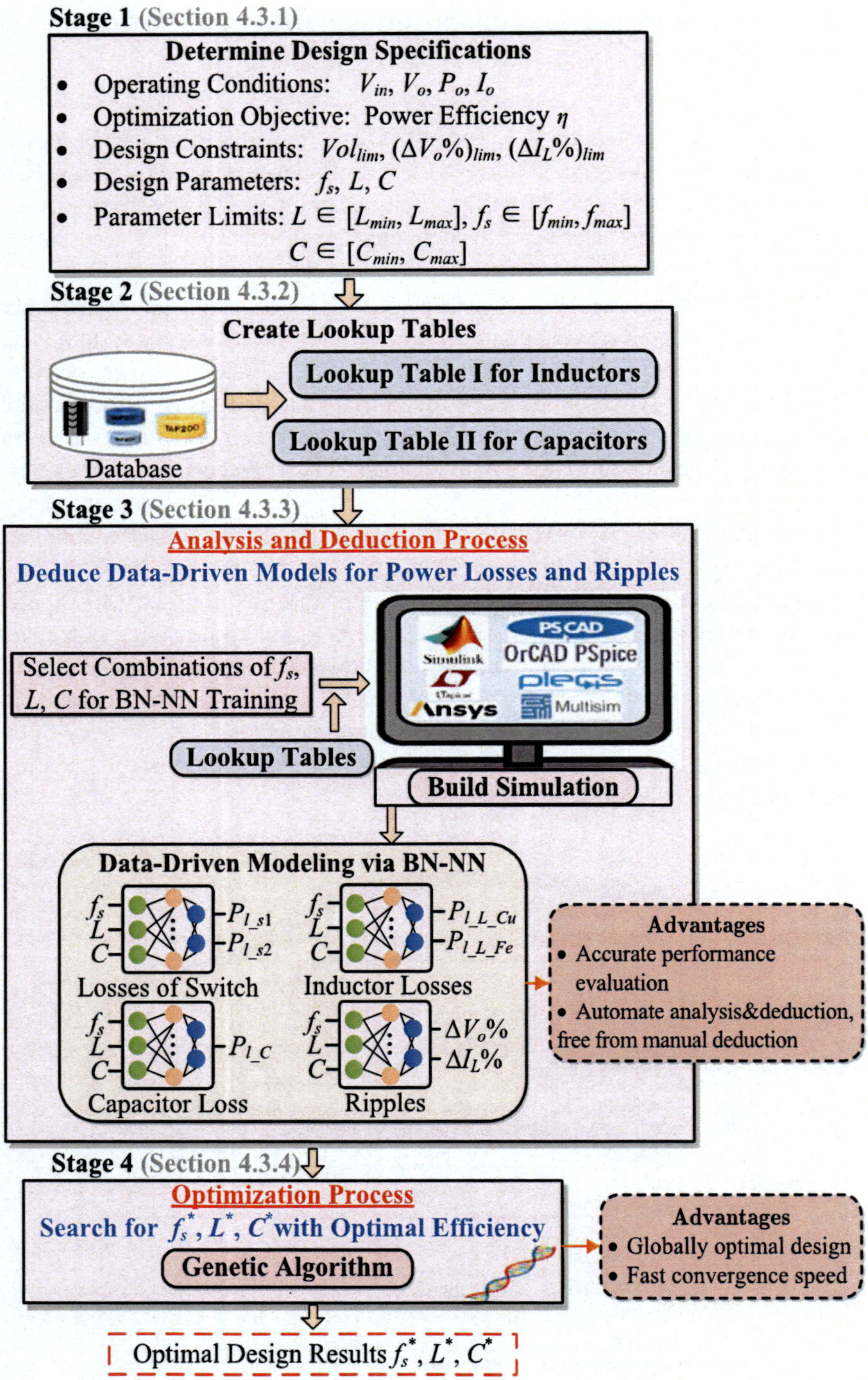

Fig. 4.5 Flowchart of the proposed AI-D approach applied in the parameter design of an efficiency-optimal synchronous Buck converter in EV

4.3.1 Stage 1: Determine Design Specifications

The initial step in the design process is to determine all the design requirements.

The design conditions should be firstly figured out, including input voltage V_{in}, output voltage V_o, output power P_o and output current I_o. Power efficiency is the optimization objective, and the design constraints include volume constraint Vol_{lim}, voltage ripple constraint $(\Delta V_o\%)_{lim}$ and current ripple constraint $(\Delta I_L\%)_{lim}$. The parameters that need to be designed contain switching frequency f_s, inductance L, and capacitance C. The limits of design parameters are also necessary as $[f_{\min}, f_{\max}]$, $[L_{\min}, L_{\max}]$, and $[C_{\min}, C_{\max}]$.

The volume as a design constraint only takes the size of the inductor and capacitor (Vol_L, Vol_C) into consideration because they will be greatly influenced by the choices of design parameters [8]. The minor effects of parameter design on the volume of other parts, like PCB board, are neglected.

4.3.2 Stage 2: Create Lookup Tables for Inductors and Capacitors

Stage 2 requires the creation of lookup tables for the inductor and capacitor values since they are both design parameters. Lookup tables record all the possible values of inductors and capacitors while taking into account practical factors to close the gap between theoretical analysis and reality.

The value of inductor L for inductors is mapped to the geometrical and magnetic characteristics of the chosen core in Lookup Table I. In the case of capacitors, Lookup Table II associates the characteristics of the chosen capacitor components with the value of capacitor C.

The toroidal-shaped core depicted in Fig. 4.6 is used as an illustration. Figure 4.7 illustrates how to establish Lookup Table I for inductors. After the specification of the fill factor of the inductor core [33], K_u, with the geometric and magnetic features of cores obtained from the core database, the maximum number of turns $N_{\max\_core}$ and the maximum inductance $L_{\max\_core}$ each core can reach are computed with (4.1) and (4.2), respectively. In (4.1) and (4.2), core inner diameter ID, and nominal inductance A_L are obtained with the database, and A_W is the area of wire. Afterward, $L_{\max\_core}$ is sorted in ascending order. $L_{\max\_core}$ partitions the selection of inductor core. For instance, if L is lower than or equal to $L_{\max\_core1}$, for the sake of smaller size of the core, core 1 is selected; If L is greater than $L_{\max\_core1}$, to avoid core saturation, core 2 is selected. Subsequently, a lookup table on the choices of cores for the reachable values of inductors can be created.

$$N \le K_u \cdot \frac{\pi ID^2}{4A_W} = N_{\max\_core} \tag{4.1}$$

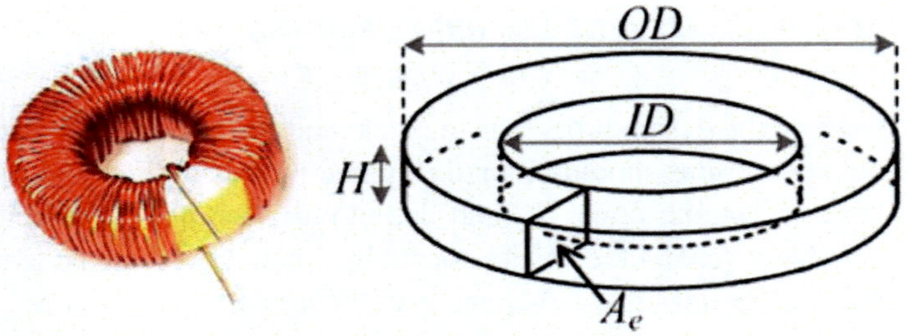

Fig. 4.6 Toroidal inductor applied in Buck converter

$$L_{\max\_core} = \frac{A_L}{1000} N^2_{\max\_core} \ \mu\text{H} \tag{4.2}$$

The flowchart shown in Fig. 4.8 is used to create a lookup table for capacitors. Possible parallel and serial connections of a single capacitor are taken into consideration in order to achieve larger capacitance values with the few available practical components. All capacitor values are available once the maximum component count for *MP*'s parallel and serial connections has been specified. Then, Lookup Table II on the various connections for various capacitor values can be constructed. The combination with the smallest volume is chosen when there are many combinations for the same planned value (e.g., for 660 μF, there can be 220 μF * 3 and 330 μF * 2).

Only toroidal cores and electrolytic capacitors are used as design samples in this case for ease of illustration. If other types of cores and capacitors are taken into account, just small alterations are needed: different types of cores and capacitors will have separate lookup tables since one lookup table will be produced for each type of core or capacitor. The types of cores and capacitors can be incorporated as the design parameters to be optimized in the AI-D process thanks to the various lookup tables that have been established and store the knowledge about various types of cores and capacitors.

4.3.3 Stage 3: Build Data-Driven Models for Power Losses, Voltage Ripple and Current Ripple

Stage 3 of the proposed AI-D technique uses simulations and BN-NN to automatically construct data-driven models of power losses, voltage ripples, and current ripples. This process assesses the power loss and ripple performance of various designs using simulation, which can provide an accurate performance rating. These simulation results will be used to train BN-NN, which will then be used to create precise data-driven models for power losses and ripples. Stage 3's detailed flowchart, which consists of 3 steps, is shown in Fig. 4.9.

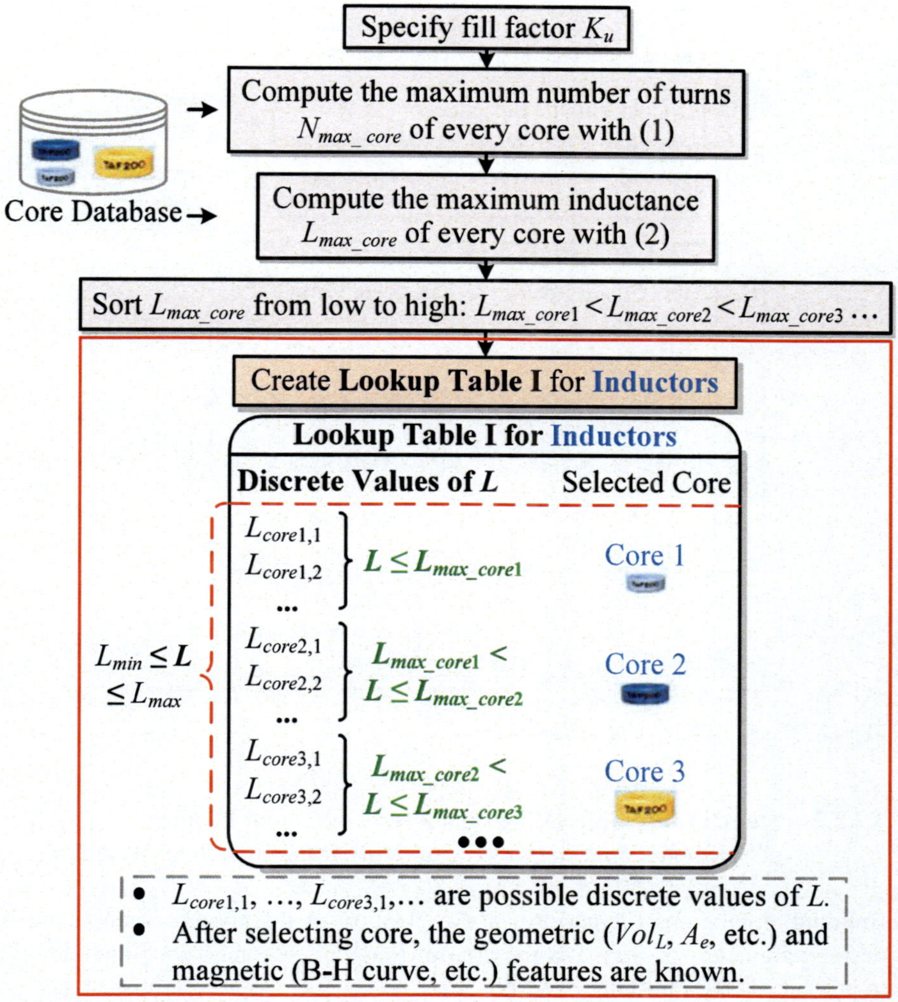

Fig. 4.7 Create Lookup Table I for inductors

4.3.3.1 Step 1: Select Combinations of f_s, L, C for BN-NN Training

In Step 1 of Stage 3, combinations of design parameters f_s, L, C should be first selected. f_s, L and C are uniformly selected within $[f_{\min}, f_{\max}]$, $[L_{\min}, L_{\max}]$ and $[C_{\min}, C_{\max}]$ for N_1, N_2, and N_3 number of points, respectively. As a result, the total number of combinations of f_s, L, C generated is $N_1 \times N_2 \times N_3$.

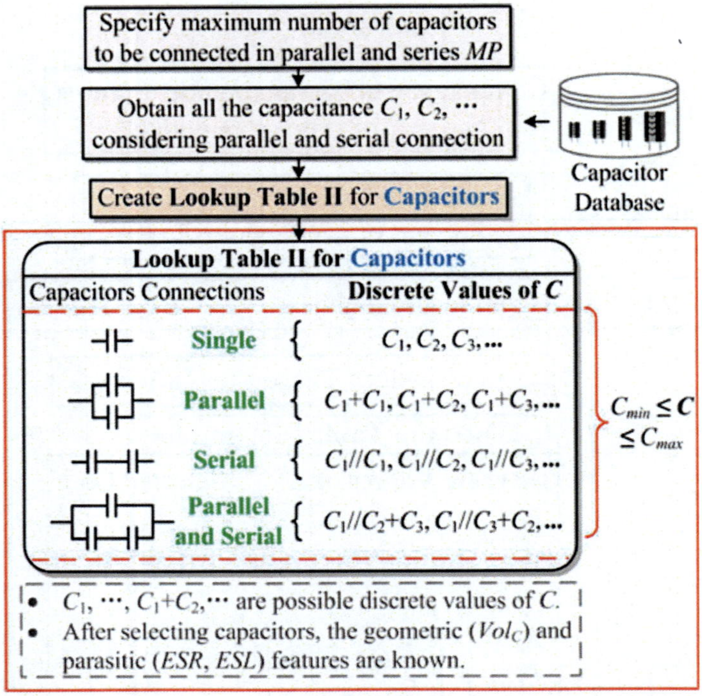

Fig. 4.8 Create Lookup Table II for capacitors

4.3.3.2 Step 2: Build Simulation for Power Losses and Ripples of Selected Combinations to Generate Data for BN-NN Training

Building simulations for performance evaluation of the chosen combinations of design parameters in Step 1 is the goal of Stage 3's second step, which attempts to produce data for training BN-NN.

Power losses, voltage ripple, current ripple, and volume are the performance indicators that require evaluation. Magnetic and electrical simulations will be used to investigate power losses among them. The electrical simulation will be used to examine voltage ripple and current ripple. And the lookup tables will be used to determine volume.

To build the magnetic simulation, the features of the inductor core (number of turns N, core geometry OD, ID, H, Ae, inductor volume Vol_L, and magnetic properties μ_i, B_{sat}) can be obtained from Lookup Table I.

To construct the electrical simulation, additional parasitic parameters, including the inductor's R_L and the capacitor's ESR and ESL, are needed. The equivalent resistance of inductor R_L can be calculated using Lookup Table I for inductors using the formula (4.3), where r is the resistance of wire per unit length. With Lookup Table II for capacitors, the equivalent series resistance ESR and the equivalent series inductance ESL of capacitor can be computed with (4.4) and (4.5) [34], respectively, where

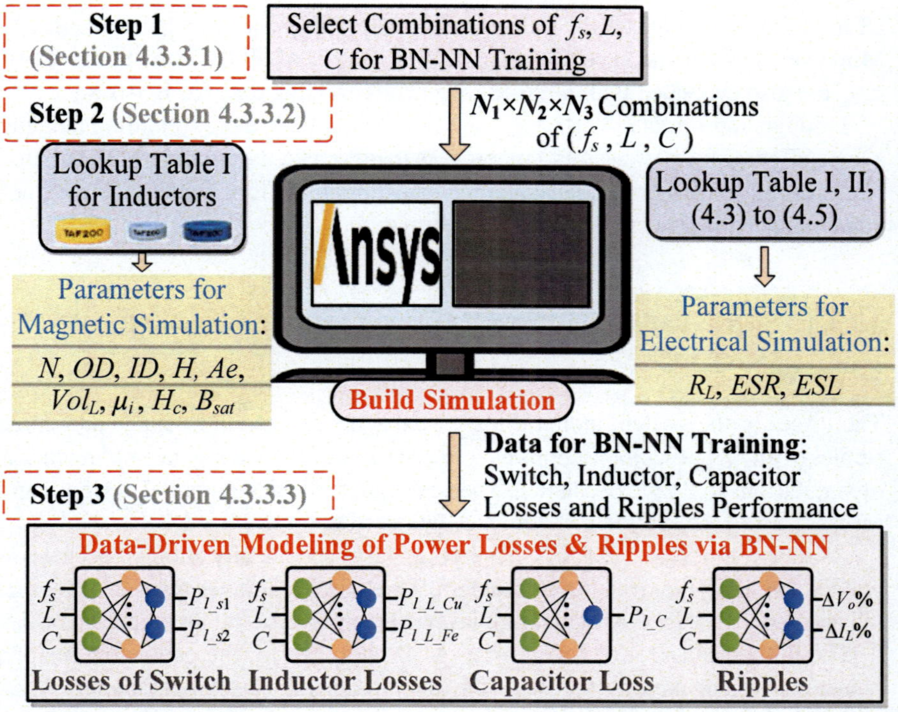

Fig. 4.9 Detailed realization of Stage 3: build data-driven models for power losses and ripples with simulations and BN-NN

$\tan\delta$ and k_{esl} are the dissipation factor and ESL factor, both of which are obtained with the capacitor database.

$$R_L = N \cdot (OD - ID + 2H) \cdot r \tag{4.3}$$

$$ESR = \frac{\tan\delta}{2\pi f_s C} \tag{4.4}$$

$$ESL = k_{esl}/C \tag{4.5}$$

The magnetic simulation tool, Ansys, is used to evaluate magnetic core loss $P_{l\_L\_Fe}$. With necessary features of the inductor core provided by Lookup Table I, magnetic simulation models can be built to evaluate $P_{l\_L\_Fe}$ of all the selected combinations of f_s, L, C.

The electrical simulation tool, LTspice, is utilized to evaluate electrical losses, voltage ripple, and current ripple. With the Spice model of power switches provided by the manufacturer, LTspice can evaluate the losses of high-side switch P_{l_s1} and low-side switch P_{l_s2}, in which the switching and conduction losses are included. Besides, with the essential parasitic parameters obtained, electrical simulation by

LTspice can also evaluate the inductor copper loss $P_{l\_L\_Cu}$, and capacitor loss P_{l_C}. Moreover, voltage and current ripples $\Delta V_o\%$, $\Delta I_L\%$ of all the combinations of f_s, L, C are also evaluated with LTspice to generate training data for BN-NN.

Through appropriate interfaces of the simulation tools, a programming language, such as Python, can be used in this process to automate the running of simulations. As a result, the programming language can run simulations, undertake parameter adjustments, and gather performance information.

4.3.3.3 Step 3: Build Data-Driven Models of Power Losses and Ripples via BN-NN

The losses of the switch, inductor losses, capacitor losses, and voltage and current ripples of all f_s, L, C combinations have been evaluated following the implementation of simulations in Step 2 of Stage 3. The performance of any potential designs can be assessed using the limited performance data used to train BN-NN.

As stated in Sect. 4.2.2.1, BN-NN in Fig. 4.3 is specifically adopted because it is good at avoiding the over-fitting problem. Three different layer types are present in the accepted BN-NN: a scalarization layer, hidden layers, and an output layer.

The initial scalarization layer of the AI-D approach for the parameter design of power converters attempts to rescale the magnitude of the design parameters f_s, L, C within the range of [0, 1]. For instance, in the original state, L is less than millihenry and f_s is greater than tens of kilohertz. To train the BN-NN objectively, they will both be rescaled into the range [0, 1].

It is presumable that there are H hidden layers, with Nh neurons in each layer. A batch-normalization layer comes after a completely linked layer in every hidden layer. The over-fitting issue, which can substantially impair NN accuracy, is something that the batch-normalization layer seeks to avoid.

The last layer is the output layer, and the outputs include losses P_{l_s1}, P_{l_s2}, $P_{l\_L\_Cu}$, $P_{l\_L\_Fe}$, P_{l_C} and ripples $\Delta V_o\%$, $\Delta I_L\%$. The complete structure of BN-NN is given in Fig. 4.10.

The simulations of all the chosen $N_1 \times N_2 \times N_3$ combinations provide performance data for power losses and ripples, which are then divided into training (70%), validating (15%), and testing (15%) sets, respectively. These sets are used to train BN-NN, choose its structure (N_h, H), and test BN-NN in the novel, previously untested designs. Different N_h and H alternatives are tried using the training and validating sets; among them, the particular N_h and H that achieve the least error on the validation set are the chosen BN-NN structure.

Until here, the automatic analysis and deduction process has been finished.

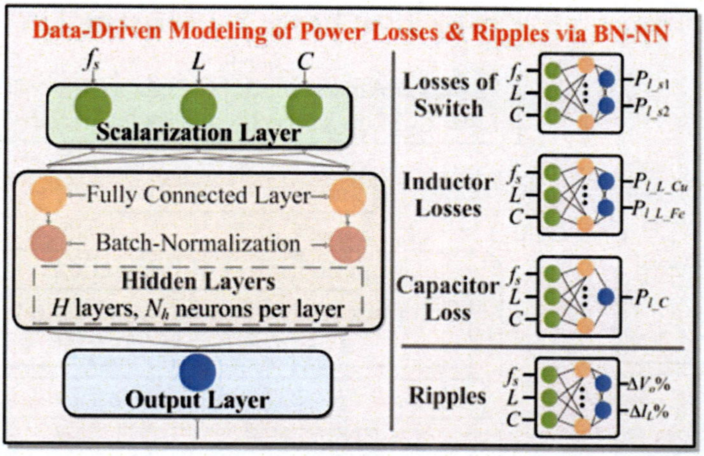

Fig. 4.10 Data-driven modeling of power losses and ripples via BN-NN

4.3.4 Stage 4: Search for Optimal Design Parameters f_s^*, L^*, C^* via Genetic Algorithm

In this stage, GA is used to discover the best f_s, L, C design parameters for power converters under the design constraints of volume, current ripple, and voltage. The optimization issue is stated as (4.6).

$$\begin{aligned}
&\min_{f_s,L,C}\left(P_{l_s1} + P_{l_s2} + P_{l_L_Cu} + P_{l_L_Fe} + P_{l_C}\right)\\
&\text{subject to:}\quad Vol_L + Vol_C \le Vol_{lim},\\
&\qquad\qquad\quad \Delta V_o\% \le (\Delta V_o\%)_{lim},\\
&\qquad\qquad\quad \Delta I_L\% \le (\Delta I_L\%)_{lim}.
\end{aligned} \tag{4.6}$$

The detailed flowchart of GA in solving (4.6) and searching for the globally optimal f_s^*, L^*, C^* is shown in Fig. 4.11, and briefly illustrated as follows. The fitness value of the ith individual F_i is computed with (4.7a), where O_i is the ith objective value as expressed in (4.7b), $O_{\max}$ and $O_{\min}$ are the maximum and minimum of O_i for all individuals, and ξ is a small constant. In (4.7b), P_{l_tot} is the total power loss, which is shown in (4.7c).

$$F_i = \frac{O_{\max} - O_i}{O_{\max} - O_{\min}} + \xi \tag{4.7a}$$

$$O_i = P_{l\_tot} + \max\left(0, \frac{Vol}{Vol_{lim}} - 1\right)$$

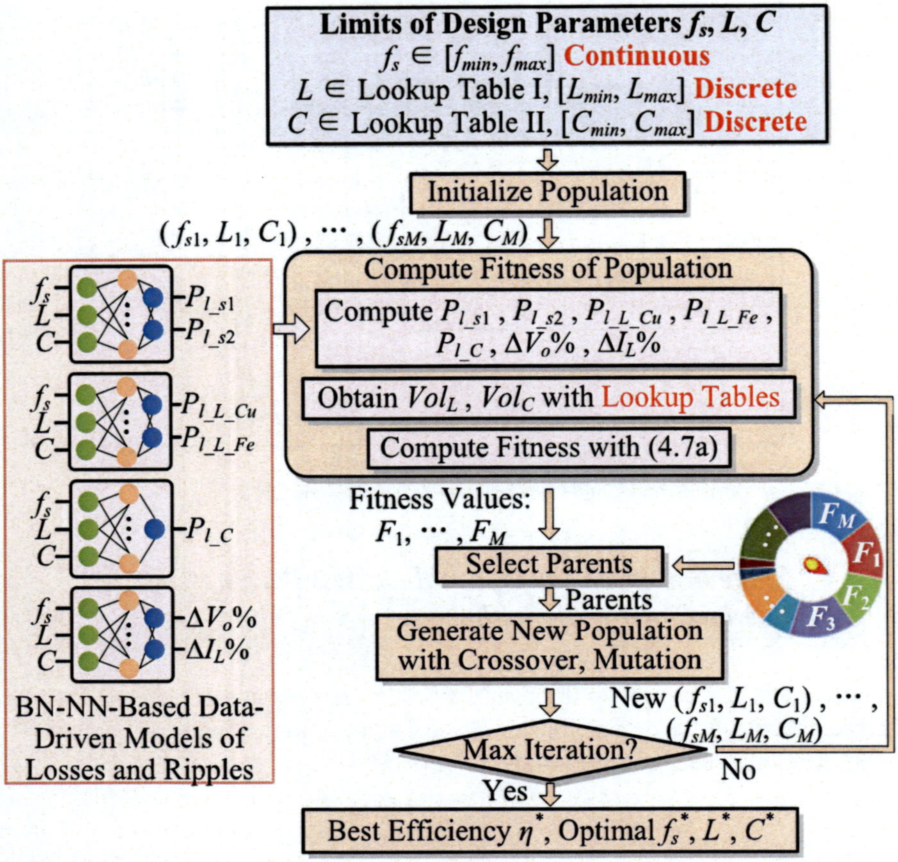

Fig. 4.11 Flowchart of GA in searching for the globally optimal $f_s{}^*$, L^*, C^*

$$+\max\left(0, \frac{\Delta V_o\%}{(\Delta V_o\%)_{lim}} - 1\right) + \max\left(0, \frac{\Delta I_L\%}{(\Delta I_L\%)_{lim}} - 1\right) \tag{4.7b}$$

$$P_{l\_tot} = P_{l\_s1} + P_{l\_s2} + P_{l\_L\_Cu} + P_{l\_L\_Fe} + P_{l\_C} \tag{4.7c}$$

To restrict the volume and ripples of the designed converters, the objective value O_i in (4.7b) has introduced penalty terms such as "max(0, $Vol/Vol_{lim} - 1$)", so that the negative effects of these constraints being exceeded are considered.

With GA's help, the optimization process has been completed up until this point. As a result, the proposed AI-based design approach's steps have all been completed, and the best synchronous Buck converter with satisfactory overall performance has been created.

4.4 Design Case of the Proposed AI-D Approach to Design an Efficiency-Optimal Synchronous Buck Converter in EV

With the proposed AI-D approach elaborated in Sect. 4.3, an efficiency-optimal synchronous Buck converter applied in 48–12 V accessory-load power supply system of EV [31] is designed. The design case is illustrated below stage by stage.

4.4.1 Determine Design Specifications

Stage 1 specifies the design parameters and specifications for the synchronous Buck converter used in the 48–12 V accessory-load power supply system of the EV, as shown in Table 4.1. Rated power P_o is selected as 100 W for the accessory loads in the EV power supply system [32, 35].

When the limits of design parameters are determined, switching frequency f_s is selected within the suitable range [20, 200 kHz] according to [36, 37]. As determined

Table 4.1 Design specifications

Operating specifications			
Topology	Synchronous Buck	V_o	12 V
V_{in}	48 V	P_o	100 W
Power switches			
Switches	IRFB4310PbF, Infineon	Dead time	200 ns
$R_{DS(on)}$	5.6 mΩ	V_{DSS}	100 V
Output *LC* filter			
Inductor cores	Toroidal TAF-200 series		
Inductor wire	UEFN/U 1 mm		
Capacitors	25 V Nippon KZE series		
Design parameters			
Switching frequency f_s	Inductance L	Capacitance C	
Limits of design parameters			
f_s	$f_{\min} = 20$ kHz; $f_{\max} = 200$ kHz		
L	$L_{\min} = 30$ μH; $L_{\max} = 2$ mH		
C	$C_{\min} = 20$ μF; $C_{\max} = 1000$ μF		
Design constraints			
Volume *Vol*	$\leq Vol_{lim} = 7$ cm^3		
Voltage ripple $\Delta V_o\%$	$\leq (\Delta V_o\%)_{lim} = 1\%$		
Current ripple $\Delta I_L\%$	$\leq (\Delta I_L\%)_{lim} = 10\%$		

by the range of f_s, the ranges of L and C are chosen from several tens of μH, μF to mH, mF, where the selection of upper limits $L_{\max}$ and $C_{\max}$ is out of cost and power density perspectives [25], and the selection of lower limits $L_{\min}$ and $C_{\min}$ considers filtering performance [8]. If a specific application requires different ranges of f_s, L and C from the given ranges in this design case, the proposed AI-D approach can still be applied with no changes in any steps.

Infineon IRFB4310PbF is chosen as the power switch due to its lower drain-source on-resistance and adequate drain-source breakdown voltage under the specified operating conditions. Other power switches can be taken into account while still using the proposed AI-D methodology. It should be noted that the Stage 3 simulations must incorporate the models of the actual power switches.

4.4.2 Create Lookup Tables for Inductors and Capacitors

Stage 2 involves creating lookup tables for capacitors and inductors. Four toroidal cores (T80-75-200, T106-75-200, T131-75-200, T150-75-200) from the TAF-200 series are used in this design instance. By following the flowchart in Fig. 4.7, Lookup Table I for inductors is created, as shown in Figs. 4.12 and 4.13, in which the fill factor K_u of inductor core is kept below 0.35 for easy manufacture [38]. The practical characteristics of the chosen cores are detailed in Fig. 4.12, and the potential discrete values of L and associated core choices are shown in Fig. 4.13.

Then, using the flowchart in Fig. 5, select the maximum number of capacitors for parallel and serial connections MP as 5. 4.8, the capacitors' Lookup Table II is formed as depicted in Figs. 4.14 and 4.15. Fig. The characteristics of a few capacitors

Lookup Table I for Inductors: Features of Selected Cores

	T80-75-200	T106-75-200	T131-75-200	T150-75-200
OD	20.2 mm	26.9 mm	33 mm	38.4 mm
ID	12.6 mm	14.5 mm	16.3 mm	21.5 mm
H	6.35 mm	11.1 mm	11.1 mm	11.1 mm
Ae	24.1 mm^2	68.8 mm^2	92.7 mm^2	93.8 mm^2
Vol_L	2.04 cm^3	6.31 cm^3	9.49 cm^3	12.9 cm^3
A_L	46	93	116	96
N_{max_core}	56	74	93	162
L_{max_core}	144μH	509μH	1003μH	2519μH

L: L_{min} — 144μH — 509μH — 1003μH — L_{max}

/* 144μH, 509μH, ... are the reachable maximum inductance of each core L_{max_core}, as computed by 0.35 fill factor */

Fig. 4.12 Lookup Table I for inductors: features of the selected cores

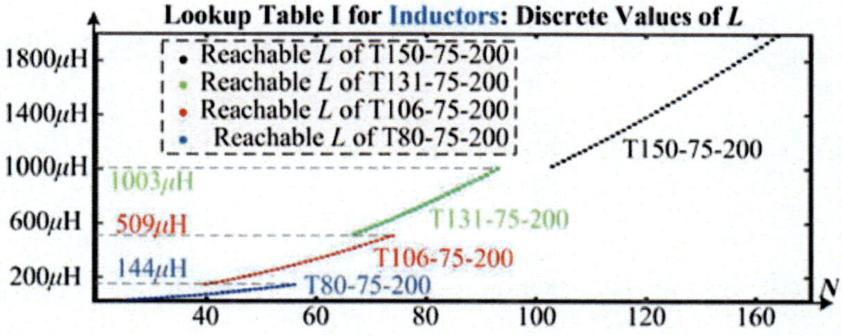

Fig. 4.13 Lookup Table I for inductors: discrete values of L

Lookup Table II for Capacitors: Features of Selected Capacitors

Dissipation factor $\tan\delta$: 0.14 ESL factor k_{esl} : 4.389×10^{-11}

	C_1	C_2	C_3	C_4	C_5	C_6	
Capacitors	27μF	47μF	56μF	100μF	220μF	330μF	•••
Vol_C **(cm³)**	0.137	0.216	0.218	0.343	0.578	0.754	

Fig. 4.14 Lookup Table II for capacitors: features of selected capacitors

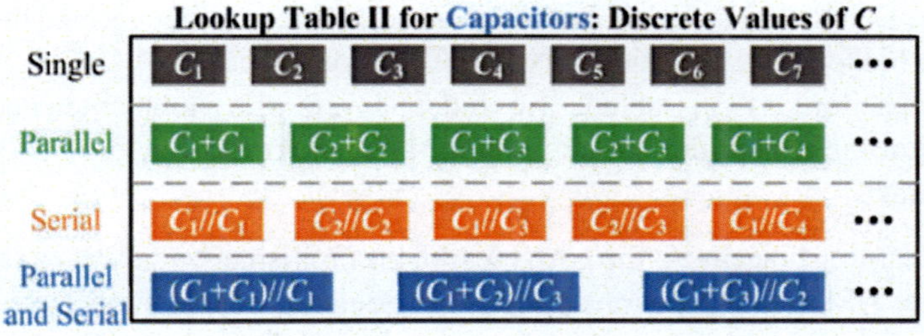

Fig. 4.15 Lookup Table II for capacitors: discrete values of C

are shown in Fig. The various discrete values of C taking into account parallel and serial connections are described in Fig. 4.15.

4.4.3 Build Data-Driven Models for Power Losses and Ripples

By using Fig. 4.9 as a guide, data-driven models for power losses and ripples with BN-NN are created in Stage 3. The specific three stages of Stage 3 are shown as follows.

Table 4.2 Mean-square-error of the compared regression methods

	Training set	Validating set	Testing set
Ridge regression	0.719	0.689	0.712
Support vector regression	0.472	0.441	0.433
Bayesian regression	0.310	0.315	0.294
NN via Matlab toolbox	0.084	0.085	0.076
Proposed BN-NN	**0.022**	**0.018**	**0.025**

- In Step 1, N_1, N_2 and N_3 in Fig. 4.9 are set to 20, 20, and 20, respectively, within the range of f_s, L and C indicated in Table 4.1, resulting in a total of 8000 permutations of design parameters being created when 202,020 = 8000 is multiplied by the number of f_s.
- In Step 2, buildings for electrical and magnetic simulation models in LTspice and Ansys, respectively All 8000 permutations are simulated to determine the power losses and ripples, which are used as training data for the BN-NN.
- In Step 3, four BN-NN are constructed: 3 hidden layers with 10 neurons each make up the BN-NN for switch losses (P_{l_s1}, P_{l_s2}); 20 neurons are located in each of the three hidden layers of the BN-NN for inductor losses ($P_{l\_L\_Cu}$, $P_{l\_L\_Fe}$); BN-NN for capacitor loss P_{l_C} includes two hidden layers with a total of ten neurons each; Each of the two hidden layers in the BN-NN for ripples ($\Delta V_o\%$, $\Delta I_L\%$) has 10 neurons.

The learning objective is the power loss of the high-side switch P_{l_s1} as an example, and many AI-based regression approaches (ridge regression, support vector regression, Bayesian regression, deep NN through Matlab Toolbox) are compared with it to highlight the greater generalization accuracy of BN-NN. The proposed BN-NN in Fig. 4.10 exhibits the least error on all the datasets, suggesting the maximum generalization accuracy across all the approaches examined, as indicated in Table 4.2.

4.4.4 Search for Optimal f_s^*, L^*, C^* via GA

Globally ideal f_s^*, L^*, C^* are discovered using GA in Stage 4 (Fig. 4.11). The best f_s^*, L^*, C^* are sought using the trained data-driven models using BN-NN for power losses and ripples and Lookup Table I and II for volume. The continuous space [f_{min}, f_{max}] is searched for fs*, L is selected from Lookup Table I, which is located within [L_{min}, L_{max}], and discrete C is selected from Lookup Table II, which is located within [C_{min}, C_{max}].

Table 4.3 Designed converter in the 48–12 V accessory-load power supply system of EV via the proposed AI-D approach

Designed efficiency-optimal synchronous Buck converter in EV	
Topology	Synchronous Buck
Switch	IRFB4310PbF, Infineon
f_s	36.6 kHz
L	281.3 μH, core T106-75-200, wire UEFN/U 1 mm, number of turns 55
C	112 μF, 2 number of 56 μF of 25 V Nippon KZE in parallel

Table 4.4 Theoretical performance of the designed synchronous Buck converter

Theoretical performance of designed converter	
Efficiency η	93.85%
Volume $Vol_L + Vol_C$	6.746 cm^3
Ripples $\Delta V_o\%$ and $\Delta I_L\%$	$\Delta V_o\% = 0.573\%$; $\Delta I_L\% = 9.7\%$

The finalized optimal design results of f_s^*, L^* and C^* are 36.6 kHz, 281.3 μH and 112 μF, respectively. They provide an optimal efficiency η^* at 93.85%.

The preset volume limitation Vol_{lim} is connected to the reason why f_s^* is optimized to 36.6 kHz. The optimized f_s^* can reach a higher value if high power density is anticipated and Vol_{lim} is set at a lower value in comparison to the one specified in this design example. However, under this scenario, switching and core losses will rise, resulting in a drop in efficiency.

Table 4.3 presents the design outcomes for the proposed AI-D technique for the efficiency-optimal synchronous Buck converter in the 48–12 V accessory-load power supply system of EV. Table 4.4 provides a summary of the planned converter's theoretical performance.

4.4.5 Average CPU Execution Time for Applying the Proposed AI-D Approach in the Design Case

The average CPU execution time of the proposed AI-D approach applied in the design case, under the computer platform with Intel Xeon CPU E5-1630 @ 3.7 GHz, 16 GB RAM, and Windows 10 operating system, is given in Table 4.5 to give an understanding of the computational time needed to implement the approach. According to Table 4.5, the majority of computational efforts are directed toward performing simulations to gather data on power loss and ripple performance, with the CPU execution time of other phases being negligible.

The average CPU execution time of the proposed AI-D approach applied in the design case, under the computer platform with Intel Xeon CPU E5-1630 @ 3.7 GHz,

Table 4.5 Average CPU execution time of AI-D approach

Stages in AI-D	Average CPU execution time
Stage 2	Total 0.232 s
Stage 3: Step 1	Total 0.074 s
Stage 3: Step 2	**Total 2 days and 4 h to run the required simulations using four CPU cores**
Stage 3: Step 3	1 min 13.4 s to train all NNs
Stage 4	Total 24.38 s

Bold terms represents as to emphasize that this step (**Stage 3: Step 2**) accounts for the majority of CPU execution time

16 GB RAM, and Windows 10 operating system, is given in Table 4.5 to give an understanding of the computational time needed to implement the approach. According to Table 4.5, the majority of computational efforts are directed toward performing simulations to gather data on power loss and ripple performance, with the CPU execution time of other phases being negligible.

4.5 Design Case of the Proposed AI-D Approach to Design an Efficiency-Optimal Synchronous Buck Converter in EV

In this part, a prototype has been created and hardware experiments have been carried out to further validate the intended synchronous Buck converter in Table 4.3 for the 48–12 V accessory-load power supply system of EV. In Fig. 4.16, the hardware platform is displayed.

Fig. 4.16 The hardware platform of the designed DC-DC converter

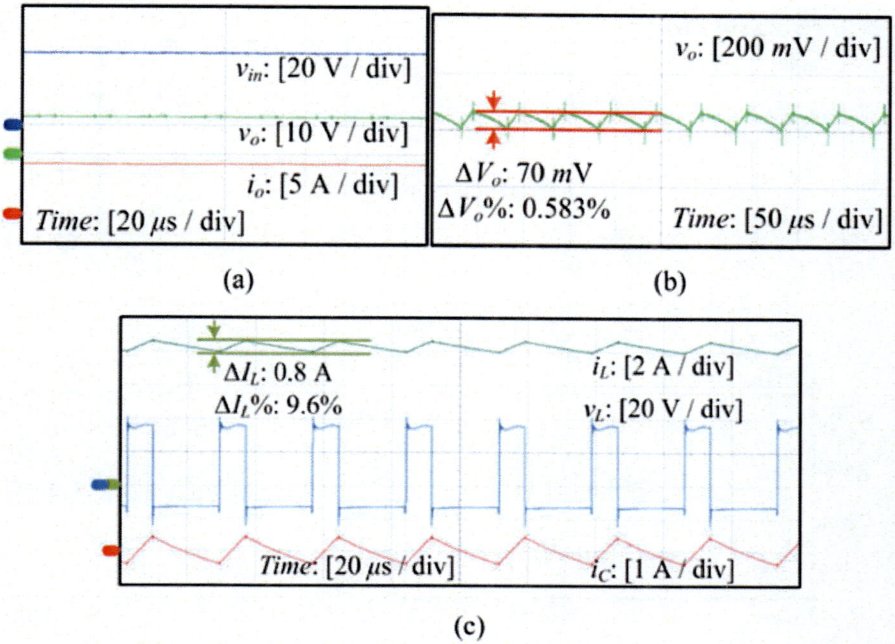

Fig. 4.17 Steady-state waveforms of the designed efficiency-optimal converter: **a** v_{in}, v_o, i_o; **b** zoom-in view of v_o; **c** v_L, i_L, i_C

4.5.1 *Steady-State Waveforms of the Designed Optimally Efficient Synchronous Buck Converter*

The waveforms of the intended converter in steady state are displayed in Fig. 4.17 under the rated circumstances listed in Table 4.1, and their notation and direction are presented in Fig. 4.4.

4.5.2 *Experimental Efficiency of the Designed Converter*

4.5.2.1 Validation of the Optimal Efficiency of the Designed Synchronous Buck Converter When f_s, *L*, *C* Are Varying

The next experiments are conducted to confirm that the converter's design achieves maximum efficiency. The switching frequency f_s varies around the optimal frequency f_s^*, as shown in Fig. 4.18a, where the efficiency reaches the highest at the designed optimal f_s^*. In addition, when L varies around the optimal L^*, as shown in Fig. 4.18b, the values smaller than L^* should be avoided because they fail to meet the required 10% current ripple constraint in Eq. (4.6), even though they achieve higher efficiency. Among the L values no smaller than L^*, the selected value L^* displays the best

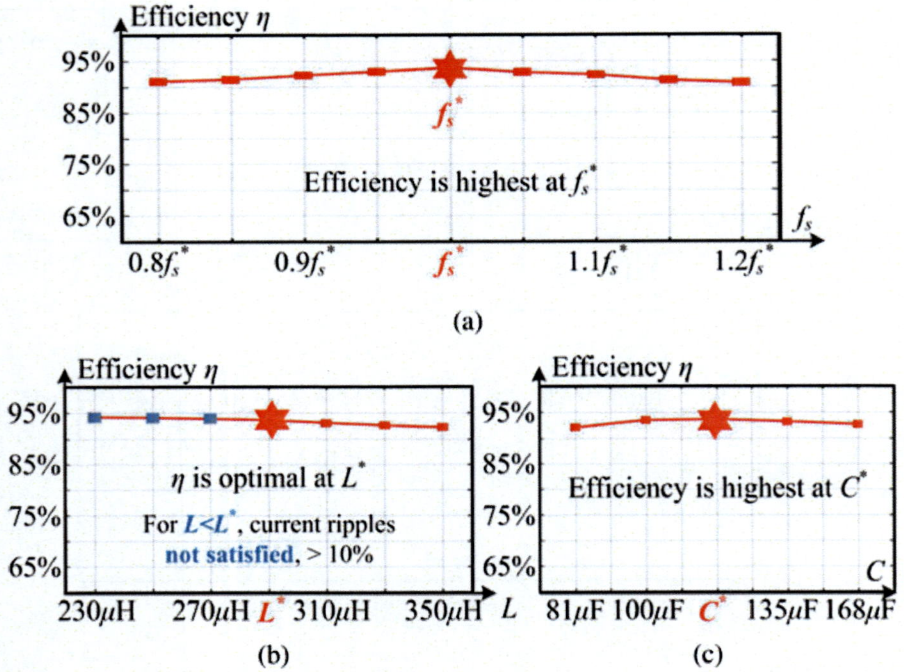

Fig. 4.18 Validation of the optimal efficiency of the designed synchronous Buck converter: **a** f_s varies within [0.8 fs^*, 1.2 fs^*]; **b** L varies within [230, 350 μH]; **c** C varies within [81, 168 μF]

efficiency, so the optimal efficiency under given constraints is still achieved at the designed optimal L^*. Besides, as shown in Fig. 4.18c, C varies within [81, 168 μF], and the results verify that the optimal efficiency at the designed C^* is achieved. Consequently, based on Fig. 4.18a–c, the designed synchronous Buck converter applied in EV enjoys optimal efficiency.

4.5.2.2 Efficiency Comparison of Designed Synchronous Buck Converters via the Proposed AI-D Approach, the CAO Approach and the Conventional Approach

To validate the superiority of the proposed AI-D approach, the conventional approach [34] and a CAO approach [8] are compared with it. Conventionally, voltage and current ripple constraints are used to determine the values of L and C [34], which are computed with (4.8a, 4.8b), where V_o, I_o, f_s, $(\Delta V_o\%)_{lim}$ and $(\Delta I_L\%)_{lim}$ are given in Table 4.1. And the computed L and C are listed in Table 4.6. As detailly discussed in [8], the compared CAO approach manually analyzes and deduces the expressions of total power loss, and then optimizes the power loss expressions for optimally efficient synchronous Buck converter. The designed L and C via the compared CAO are shown in Table 4.7.

$$L = \frac{(1 - D) \cdot V_o}{f_s \cdot (\Delta I_L\%)_{lim} \cdot I_o} \tag{4.8a}$$

$$C = \frac{(\pi + 4 \tan \delta) \cdot (\Delta I_L\%)_{lim} \cdot I_o}{8\pi f_s \cdot V_o \cdot (\Delta V_o\%)_{lim}} \tag{4.8b}$$

Figure 4.19 displays a comparison of experimental effectiveness. Evidently, the converter developed using the proposed AI-D technique outperforms the ones developed conventionally and using the CAO approach at various load levels. While the traditionally constructed converter in Table 4.6 achieves a peak efficiency of just 90.16%, the AI-D converter in Table 4.3 gets a peak efficiency of 93.68%. The efficiency of the design using the proposed AI-D is 1.22% greater than the 92.46% attained by the design via CAO in Table 4.7.

Table 4.6 Conventionally designed synchronous Buck converter

Conventionally designed synchronous Buck converter	
f_s	20 kHz
L	540 μH, core T135-75-200, wire UEFN/U 1 mm, number of turns 68
C	56 μF, single 56 μF of 25 V Nippon KZE

Table 4.7 CAO-designed synchronous Buck converter

CAO-designed synchronous Buck converter	
f_s	36.6 kHz
L	334.8 μH, core T106-75-200, wire UEFN/U 1 mm, number of turns 60
C	94 μF, 2 number of 47 μF of 25 V Nippon KZE in parallel

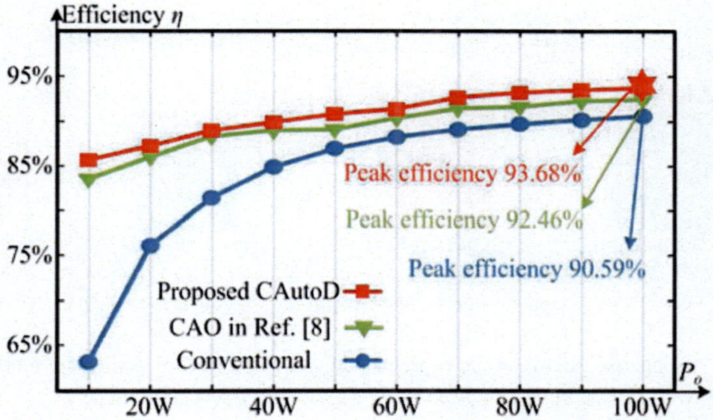

Fig. 4.19 Experimental efficiency comparison of designed converters via the proposed AI-D approach, the CAO approach and the conventional approach

4.5.3 Experimental Volume and Ripples of the Designed Converter

The AI-D technique was used to build the converter, and the volume of the inductor and capacitor is measured to be 6.9 cm^3, which satisfies the 7 cm^3 volume restriction.

As for the experimental ripples, based on Fig. 4.17, the experimental voltage ripple $\Delta V_o\%$ and current ripple $\Delta I_L\%$ are 0.583% and 9.6%, respectively, both of which meet the corresponding constraints at 1% and 10%.

4.5.4 Comparison Between the Experimental and Theoretical Efficiency, Volume and Ripples of the Designed Converter

In this section, the theoretical performance in Table 4.4 is contrasted with the experimental power losses, volume, and ripples. The practical performance is nearly identical to that in theory, and the average error is just 1.31%, as shown in Fig. 4.20. Thanks to the specifically adopted BN-NN in the analysis and deduction process and the adopted GA in the optimization process, this demonstrates the viability and high accuracy of the proposed AI-D technique.

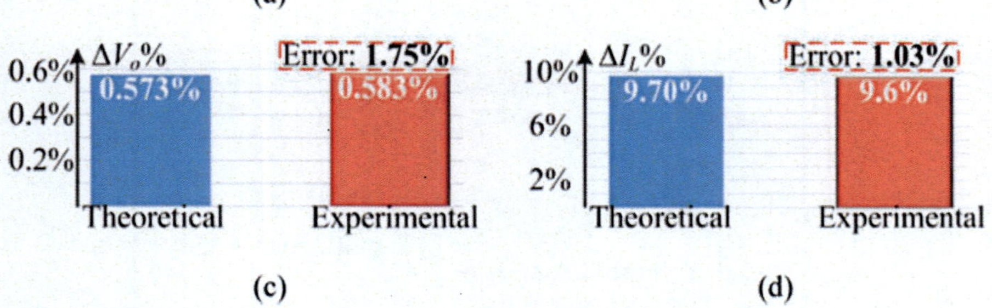

Fig. 4.20 Comparison between the experimental and theoretical performance: **a** efficiency η; **b** volume $Vol_C + Vol_L$; **c** voltage ripple $\Delta V_o\%$; **d** current ripple $\Delta I_L\%$

4.6 Conclusion

In this chapter, a parameter design technique for power converters based on artificial intelligence (AI-D) is proposed. The analysis and deduction process as well as the optimization process can both be carried out automatically using the proposed AI-D technique. It has two very noteworthy benefits: In addition to reducing the workload for engineers and realizing a quick and simple design process, it also makes it possible to guarantee excellent design correctness due to the reduced human dependency.

The batch-normalization neural network (BN-NN) and simulation tools are used in the proposed AI-D technique to develop data-driven models for the optimization targets and design constraints, achieving automation in the analysis and deduction process. Because it is good at preventing over-fitting issues, the specifically applied BN-NN is advantageous for design correctness. Additionally, a genetic algorithm is employed to look for the best design solutions to automate the optimization process.

The parameter design of an efficiency-oriented synchronous Buck converter in the 48–12 V accessory-load supply system in EV validates the proposed AI-D technique. The proposed AI-D approach's viability and great accuracy have also been confirmed by hardware trials.

References

1. Z. Gong *et al.*, 'Distributed Control of Active Cell Balancing and Low-Voltage Bus Regulation in Electric Vehicles Using Hierarchical Model-Predictive Control', *IEEE Trans. Ind. Electron.*, vol. 67, no. 12, pp. 10464–10473, Dec. 2020.
2. F. Blaabjerg, Z. Chen, and S. B. Kjaer, 'Power Electronics as Efficient Interface in Dispersed Power Generation Systems', *IEEE Trans. Power Electron.*, vol. 19, no. 5, pp. 1184–1194, Sep. 2004.
3. Y. Cao and J. A. A. Qahouq, 'Analysis and Evaluation of a Dual-Variable Closed-Loop Control of Power Converter With Wireless and Nonwireless Power Transfer', *IEEE Trans. Ind. Electron.*, vol. 66, no. 4, pp. 2668–2679, Apr. 2019.
4. F. Lin, X. G. Zhang, and X. Li, 'Design Methodology for Symmetric CLLC Resonant DC Transformer Considering Voltage Conversion Ratio, System Stability and Efficiency', *IEEE Trans. Power Electron.*, pp. 1–1, 2021.
5. Z. Amjadi and S. S. Williamson, 'Power-Electronics-Based Solutions for Plug-in Hybrid Electric Vehicle Energy Storage and Management Systems', *IEEE Trans. Ind. Electron.*, vol. 57, no. 2, pp. 608–616, Feb. 2010.
6. P. Lin, C. Zhang, X. Zhang, H. H. C. Iu, Y. Yang, and F. Blaabjerg, 'Finite-Time Large Signal Stabilization for High Power DC Microgrids With Exact Offsetting of Destabilizing Effects', *IEEE Trans. Ind. Electron.*, vol. 68, no. 5, pp. 4014–4026, May 2021.
7. F. Blaabjerg, H. Wang, I. Vernica, B. Liu, and P. Davari, 'Reliability of Power Electronic Systems for EV/HEV Applications', *Proc. IEEE*, pp. 1–17, 2020.
8. X. Li, X. Zhang, and F. Lin, 'Multi-Objective Design of Output LC Filter for Buck Converter via the Coevolving-AMOSA Algorithm', *IEEE Access*, vol. 9, pp. 11884–11894, 2021.
9. T. B. Soeiro, J. Muhlethaler, J. Linner, P. Ranstad, and J. W. Kolar, 'Automated Design of a High-Power High-Frequency LCC Resonant Converter for Electrostatic Precipitators', *IEEE Trans. Ind. Electron.*, vol. 60, no. 11, pp. 4805–4819, Nov. 2013.

10. Y.-S. Hwang, J.-J. Chen, J. Yang, and Y. Ku, 'A Low-EMI Continuous-Time Delta-Sigma-Modulator Buck Converter With Transient Response Eruption Techniques', *IEEE Trans. Ind. Electron.*, vol. 67, no. 8, pp. 6854–6863, Aug. 2020.
11. Ruiyang Yu, B. M. H. Pong, B. W.-K. Ling, and J. Lam, 'Two-Stage Optimization Method for Efficient Power Converter Design Including Light Load Operation', *IEEE Trans. Power Electron.*, vol. 27, no. 3, pp. 1327–1337, Mar. 2012.
12. J. Chen, Q. Luo, J. Huang, Q. He, and X. Du, 'A Complete Switching Analytical Model of Low-Voltage eGaN HEMTs and Its Application in Loss Analysis', *IEEE Trans. Ind. Electron.*, vol. 67, no. 2, pp. 1615–1625, Feb. 2020.
13. L. F. Costa, G. Buticchi, and M. Liserre, 'Optimum Design of a Multiple-Active-Bridge DC–DC Converter for Smart Transformer', *IEEE Trans. Power Electron.*, vol. 33, no. 12, pp. 10112–10121, Dec. 2018.
14. S. E. De Leon-Aldaco, H. Calleja, and J. Aguayo Alquicira, 'Metaheuristic Optimization Methods Applied to Power Converters: A Review', *IEEE Trans. Power Electron.*, vol. 30, no. 12, pp. 6791–6803, Dec. 2015.
15. J. Song, F. Dong, J. Zhao, H. Wang, Z. He, and L. Wang, 'An Efficient Multiobjective Design Optimization Method for a PMSLM Based on an Extreme Learning Machine', *IEEE Trans. Ind. Electron.*, vol. 66, no. 2, pp. 1001–1011, Feb. 2019.
16. F. Filho, H. Z. Maia, T. H. A. Mateus, B. Ozpineci, L. M. Tolbert, and J. O. P. Pinto, 'Adaptive Selective Harmonic Minimization Based on ANNs for Cascade Multilevel Inverters With Varying DC Sources', *IEEE Trans. Ind. Electron.*, vol. 60, no. 5, pp. 1955–1962, May 2013.
17. S. Wang, T. Dragicevic, G. F. Gontijo, S. K. Chaudhary, and R. Teodorescu, 'Machine Learning Emulation of Model Predictive Control for Modular Multilevel Converters', *IEEE Trans. Ind. Electron.*, pp. 1–1, 2020.
18. F.-J. Lin, K.-C. Lu, and B.-H. Yang, 'Recurrent Fuzzy Cerebellar Model Articulation Neural Network Based Power Control of a Single-Stage Three-Phase Grid-Connected Photovoltaic System During Grid Faults', *IEEE Trans. Ind. Electron.*, vol. 64, no. 2, pp. 1258–1268, Feb. 2017.
19. F. Lin, X. Zhang, X. Li, H. Ma, and C. Cai, 'Design of Symmetrical CLLC Resonant DC Transformer Considering Voltage Transfer Ratio and Cascaded System Stability', *IEEE Trans. Ind. Electron.*, pp. 1–1, 2021.
20. Z. Gmyrek, A. Boglietti, and A. Cavagnino, 'Iron Loss Prediction With PWM Supply Using Low- and High-Frequency Measurements: Analysis and Results Comparison', *IEEE Trans. Ind. Electron.*, vol. 55, no. 4, pp. 1722–1728, Apr. 2008.
21. A. De Nardo, N. Femia, G. Petrone, and G. Spagnuolo, 'Optimal Buck Converter Output Filter Design for Point-of-Load Applications', *IEEE Transactions on Industrial Electronics*, vol. 57, no. 4, pp. 1330–1341, Apr. 2010.
22. T. C. Neugebauer and D. J. Perreault, 'Computer-aided optimization of DC/DC converters for automotive applications', *IEEE Trans. Power Electron.*, vol. 18, no. 3, pp. 775–783, May 2003.
23. B. Zhao, X. Zhang, and J. Huang, 'AI Algorithm-Based Two-Stage Optimal Design Methodology of High-Efficiency *CLLC* Resonant Converters for the Hybrid AC–DC Microgrid Applications', *IEEE Trans. Ind. Electron.*, vol. 66, no. 12, pp. 9756–9767, Dec. 2019.
24. G. Adinolfi, G. Graditi, P. Siano, and A. Piccolo, 'Multiobjective Optimal Design of Photovoltaic Synchronous Boost Converters Assessing Efficiency, Reliability, and Cost Savings', *IEEE Trans. Ind. Inf.*, vol. 11, no. 5, pp. 1038–1048, Oct. 2015.
25. Y. Liu, M. Huang, H. Wang, X. Zha, J. Gong, and J. Sun, 'Reliability-Oriented Optimization of the LC Filter in a Buck DC-DC Converter', *IEEE Transactions on Power Electronics*, vol. 32, no. 8, pp. 6323–6337, Aug. 2017.
26. W. Xie, J. Wang, C. Xing, S. Guo, M. Guo, and L. Zhu, 'Variational Autoencoder Bidirectional Long and Short-term Memory Neural Network Soft-sensor Model Based on Batch Training Strategy', *IEEE Trans. Ind. Inf.*, pp. 1–1, 2020.
27. D. Bacciu and F. Crecchi, 'Augmenting Recurrent Neural Networks Resilience by Dropout', *IEEE Trans. Neural Netw. Learning Syst.*, vol. 31, no. 1, pp. 345–351, Jan. 2020.

28. S. Wu *et al.*, '$L1$ -Norm Batch Normalization for Efficient Training of Deep Neural Networks', *IEEE Trans. Neural Netw. Learning Syst.*, vol. 30, no. 7, pp. 2043–2051, Jul. 2019.
29. X. Li, K. Mao, F. Lin, and X. Zhang, 'Particle swarm optimization with state-based adaptive velocity limit strategy', *Neurocomputing*, vol. 447, pp. 64–79, Aug. 2021.
30. R. A. Hanifah, S. F. Toha, S. Ahmad, and Mohd. K. Hassan, 'Swarm-Intelligence Tuned Current Reduction for Power-Assisted Steering Control in Electric Vehicles', *IEEE Trans. Ind. Electron.*, vol. 65, no. 9, pp. 7202–7210, Sep. 2018.
31. G. Chen, Y. Deng, J. Dong, Y. Hu, L. Jiang, and X. He, 'Integrated Multiple-Output Synchronous Buck Converter for Electric Vehicle Power Supply', *IEEE Trans. Veh. Technol.*, vol. 66, no. 7, pp. 5752–5761, Jul. 2017.
32. T. Kim and S. Kwak, 'A Flexible Voltage Bus Converter for the 48-/12-V Dual Supply System in Electrified Vehicles', *IEEE Trans. Veh. Technol.*, vol. 66, no. 3, pp. 2010–2018, Mar. 2017.
33. 'Magnetics', in *Grounding and Shielding*, Hoboken, NJ, USA: John Wiley & Sons, Inc, 2016, pp. 27–47.
34. R. W. Erickson, *Fundamentals of power electronics*. New York: Chapman & Hall, 1997.
35. K. L. Snyder, 'Advancement And Validation Of A Plug-In Hybrid Electric Vehicle Model Utilizing Experimental Data From Vehicle Testing', p. 156.
36. S. Pang *et al.*, 'Interconnection and Damping Assignment Passivity-Based Control Applied to On-Board DC–DC Power Converter System Supplying Constant Power Load', *IEEE Trans. on Ind. Applicat.*, vol. 55, no. 6, pp. 6476–6485, Nov. 2019.
37. S. Dusmez, A. Hasanzadeh, and A. Khaligh, 'Comparative Analysis of Bidirectional Three-Level DC–DC Converter for Automotive Applications', *IEEE Trans. Ind. Electron.*, vol. 62, no. 5, pp. 3305–3315, May 2015.
38. S. D. Sudhoff, Ed., 'Introduction To Inductor Design', in *Power Magnetic Devices*, Hoboken, New Jersey: John Wiley & Sons, Inc., 2014, pp. 113–132.

Chapter 5
The Proposed Efficiency-Oriented Two-Stage Optimal Design Methodology for Special Power Converter in Space Travelling-Wave Tube Amplifier Applications

5.1 Introduction

Space Traveling-Wave Tube Amplifiers (TWTAs) are widely used as vacuum electronic devices in satellite communications, navigations, electronic countermeasures, and radars, as the final-stage power amplifiers [1, 2]. A TWTA typically consists of two components: An Electronic Power Conditioner (EPC) and a Travelling-Wave Tube (TWT) [3]. The primary purpose of the EPC is to provide the TWT with the needed power supply.

In space applications, solar panels are used to generate electricity with low efficiency (for the single-crystalline single junction Si technology, the conversion efficacy is lower than 30%) [4]. Furthermore, the TWTAs consume more than 80% of the electric power in the spacecraft. As a result, the efficiency of the TWTA is critical in order to reduce the volume and weight of solar panels [5]. Because the EPC is the main power conversion component of the TWTA, its performance, such as efficiency and power density, has a significant impact on the overall performance of the TWTA system. Therefore, it is critical for the EPC to operate under high efficiency and high-power density conditions.

The two-stage power converter structure, as illustrated in Fig. 5.1, is always used in an EPC [6, 7]. The first stage is a pre-regulator that always uses a close-loop Buck or Boost converter, which has been extensively studied [8, 9]. The second stage is typically a high-frequency high-voltage *LCLC* resonant converter as an open-loop DC transformer whose functions include increasing the input voltage, providing galvanic isolation, and maintaining high efficiency. Therefore, the pre-regulator will only adjust the *LCLC* resonant converter's input voltage while the switching frequency, duty cycle, and voltage gain stay constant. However, it should be noted that the DC power supply can be used to vary the input voltage of the *LCLC* resonant converter. As a result, the pre-regulator is not mentioned in this chapter, which primarily concentrates on the second stage, the *LCLC* resonant converter.

In Fig. 5.2, the *LCLC* resonant converter is displayed. In the space TWTA applications, in order to maximize efficiency by fully recycling the energy stored in the

X. Zhang et al., *Automated Design of Electrical Converters with Advanced AI Algorithms*,
https://doi.org/10.1007/978-981-99-0459-4_5

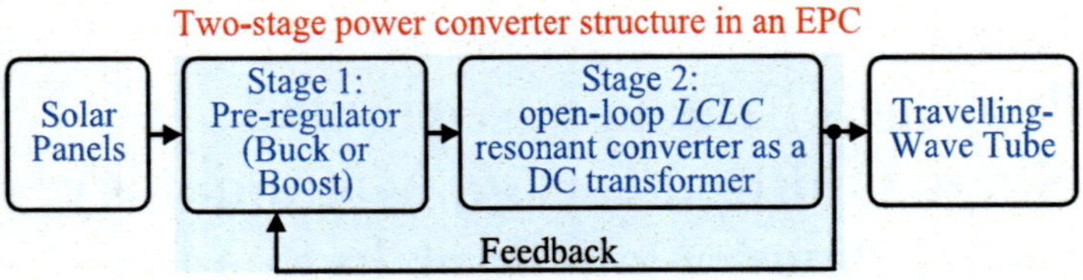

Fig. 5.1 Two-stage power converter structure in an EPC

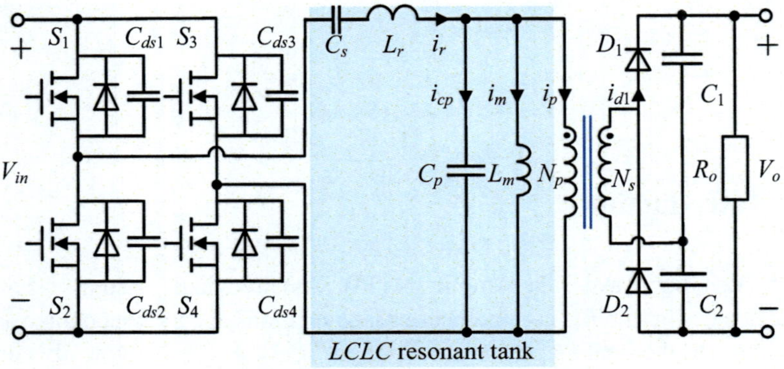

Fig. 5.2 The *LCLC* resonant converter for space TWTA applications

transformer parasitics, all the transformer parasitics are used in the resonant tank, including the leakage inductance (L_r), the magnetizing inductance (L_m), and the parasitic capacitance (C_p) [6, 7, 10]. Therefore, only an external C_s is required for the *LCLC* resonant tank.

In an EPC, research on the second stage is relatively few and more difficult than research on the first stage [3, 6]. As a result, the focus of this chapter will be on the second stage of an EPC, which is the *LCLC* resonant converter in TWTA applications.

Open loop [6, 7] and closed loop [10–18] studies on *LCLC* resonant converters can be summed up as prior research. In [6], the ZCS and ZVS *LCLC* resonant converter is proposed in the TWTA telecommunication satellite applications. The ZVS and Zero-Current-Switching (ZCS) conditions of the open-loop *LCLC* resonant converter are both analyzed in [7]. Additionally, the analytical equations of the *LCLC* resonant converter are also derived. An *LCLC* resonant converter with fixed-frequency phase-shift is proposed in [10]. In [11], an *LCLC* resonant tank is created by adding extra capacitance to the LLC resonant tank, and the traditional LLC control approach is used in the *LCLC* resonant converter. It is presented in [12] to use a unique closed loop to control the voltage gain of the *LCLC* resonant converter. The *LCLC* resonant converter control is re-designed in [13] for dual outputs and the situation when the high leakage inductance is taken into account. In addition to the switching losses that have been studied in [6, 7, 14–16] concerning the power loss analysis of the *LCLC* resonant converter, the transformer losses, comprising the winding loss and core loss, have also been analyzed in [7, 14, 18].

As previously said, because electric power is produced from solar panels in space TWTA applications, the efficiency of the *LCLC* resonant converter is critical in order to reduce the volume and weight of the solar panels. However, these chapters focused primarily on state analysis, control, soft switching, and transformer power loss, while total power loss optimization, including the conduction loss of the main switches, copper loss, core loss, and dielectric loss of the transformer, and the conduction loss of the rectifiers, were not considered [6, 7, 14–18]. In other words, previous studies were topology-oriented or control-oriented rather than total power loss-oriented. As a result, the optimization of the *LCLC* resonant converter's total power loss is not investigated. It should be noted that the switching loss accounts for only a portion of the total power loss in the *LCLC* resonant converter. In addition to the switching loss, the total power loss comprises the driving and conduction losses of the primary switches; the rectifier loss [19], copper loss [20], core loss [21], and, most importantly, the transformer's dielectric loss [22]. In other words, ZCS and ZVS indicate that the switching loss of the *LCLC* resonant converter is reduced but not the total power loss of the converter. As a result, in order to attain high efficiency, the converter's total power loss must be further adjusted. Unfortunately, to the best of the authors' knowledge, due to the many variables and mutual couplings of the *LCLC* resonant converter, the optimization of total power loss is rarely recorded.

Nowadays, the population-based metaheuristic algorithms, such as Genetic Algorithm (GA), Particle Swarm Optimization (PSO), Ant Colony algorithm (ACO), and Bee Colony algorithms (BCO), are extensively applied in the optimization problems, because of their capability to handle complicated issues and parallelism [23–31], because of the advantages of superior performance for global optimization and the ability to handle objective functions that are stationary or transient, linear or nonlinear, continuous or discontinuous. The ACO algorithm has been utilized to handle combinatorial optimization problems such as induction motor control [26], and traveling salesman problem (TSP) [27] as a universal population-based stochastic optimization method. Nevertheless, the convergence rate and the quality of the solutions are highly dependent on the original values. Furthermore, a significant amount of calculation is necessary. As another population-based search algorithm, the BCO algorithm has been applied for the Sheppard–Taylor PFC converter [28], optimal distributed generation allocation, and sizing in distribution systems [29]. Nevertheless, the BCO algorithm is a local optimum algorithm. Additionally, the computation time is still lengthy, particularly when it is applied to the solution of extensive optimization issues. The global optimum and iteration number of convergence in the GA or PSO algorithm are mutually exclusive: better worldwide optimums are always associated with higher iteration numbers, whereas inferior global optimums are always associated with lower iteration numbers [30, 31]. Furthermore, the number of iterations and the goal function are not optimized because the initial parameters and issue representation are always chosen based on user experience. In order to attain the requisite accuracy, the iteration number is sacrificed in the experience-based PSO method. Four parameters, *Weight.start, Weight.end, kind,* and *Vel.max*, to be precise, where *Weight.start* and *Weight.end* represent the beginning and ending values of the iteration velocity. The primary elements that affect the algorithm's effectiveness are inertia, *kind*, which depicts how *Weight* changes, and *Vel.max*, which is the maximum velocity. PSO will be lacking if these four parameters are chosen at random. In this chapter, a hybrid metaheuristic algorithm called GA + PSO is introduced to decrease the number of iterations and calculation time. The GA method is utilized to optimize the initial parameters for the PSO algorithm. The performance of PSO can be increased by cutting down on the number of iterations and calculation time using the GA's optimized initial parameters.

As a result, the goal of this chapter is to minimize the total power loss of the *LCLC* resonant converter while taking into account all relevant factors, including switching loss and conduction loss of the main switches, copper loss, core loss, and particularly dielectric loss of the transformer, and conduction loss of the rectifiers. The total power loss in the *LCLC* resonant converter is first determined after reviewing the calculations of the key parameters. After that, an efficiency-oriented two-stage optimal design method for the *LCLC* resonant converter is presented: in the first stage, a GA (Genetic Algorithm) + PSO (Particle Swarm Optimization) algorithm is presented to optimize the total power loss and the optimal parameters, including L_r, L_m, C_s, and C_p, are derived; in the second stage, the optimal parameters are realized by the hybrid electromagnetic analysis and a single-layer partially-interleaved transformer structure is presented. The transformer is used to construct the best *LCLC* resonant converter. Finally, simulations and experiments are used to validate the efficiency-oriented two-stage optimum design method and the presented transformer structure.

The remainder of this chapter is arranged below. The total power loss, which is the objective function of the efficiency-oriented two-stage optimal design method, is calculated from the equations of the key parameters in the *LCLC* resonant converter in Sect. 5.2. The proposed efficiency-oriented two-stage optimal design method is the major topic of Sect. 5.3. The proposed two-stage optimal design method's flowchart is first shown. The first stage, which is the extraction of the optimal parameters based on the presented GA + PSO algorithm, and the second stage, which is the realization of the optimal parameters based on the presented single-layer partially-interleaved transformer structure, are then developed successively. The presented two-stage optimal design technique and the single-layer partially-interleaved transformer construction are validated using simulations and experiments in Sect. 5.4. The entire chapter is summarized in Sect. 5.5.

5.2 Preliminary of the Proposed Efficiency-Oriented Two-Stage Optimal Design Method: Review of the LCLC Resonant Converter and the Calculation of the Total Power Loss

This section reviews the equations for the critical *LCLC* resonant converter parameters and computes the total power loss. The main parameter equations that will be utilized to calculate the total power loss are reviewed in the first section. The total power loss, which includes the driving loss and conduction loss of the main switches, the core loss, copper loss, and dielectric loss of the transformer, and the conduction loss of the rectifiers, is computed in the second section using the equations from the first part.

5.2.1 Review of the Calculations of the Main Parameters in the LCLC Resonant Converter

The typical waveforms of the *LCLC* resonant converter in the TWTA applications are depicted in Fig. 5.3, under the ZCS and ZVS circumstances. The following are the waveforms of interest:

(a) the driving signals of S_1, S_2, S_3 and S_4, $v_{gs1}(t)$, $v_{gs2}(t)$, $v_{gs3}(t)$ and $v_{gs4}(t)$;
(b) the resonant current, $i_r(t)$;
(c) the voltage across the junction capacitance of S_1 and S_4, C_{ds1} and C_{ds4}, $v_{ds1}(t)$ and $v_{ds4}(t)$;
(d) the magnetizing current, $i_m(t)$;
(e) the voltage across the parallel capacitor (C_p), $v_{cp}(t)$;
(f) the voltage across the series capacitor (C_s), $v_{cs}(t)$;
(g) the voltage across the rectifier diode D_1, $v_{d1}(t)$;
(h) the current through the rectifier diode D_1, $i_{d1}(t)$.

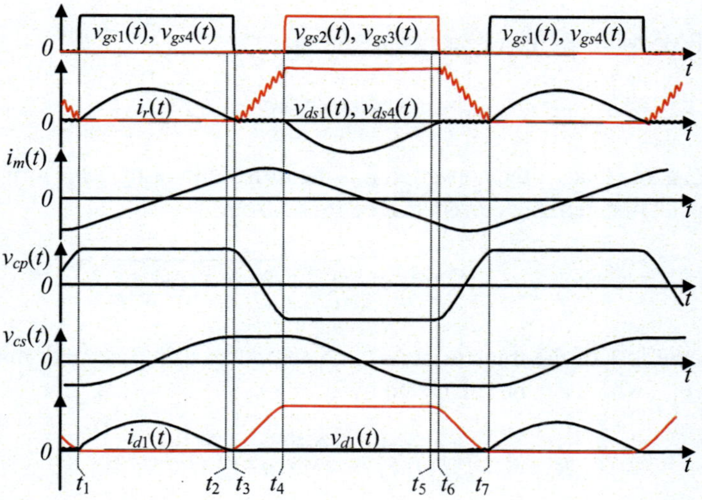

Fig. 5.3 Typical waveforms of *LCLC* resonant converters

where $\mathbf{i}_{Labc}$, $\mathbf{v}_{oabc}$, $\mathbf{i}_{oabc}$, $\boldsymbol{\mu}_{abc}$ denote the vectors of the filter inductor currents $[\, i_{La}\; i_{Lb}\; i_{Lc}\,]^T$, filter capacitor voltages $[\, v_{oa}\; v_{ob}\; v_{oc}\,]^T$, load currents $[\, i_{oa}\; i_{ob}\; i_{oc}\,]^T$ and duty cycles $[\, \mu_a\; \mu_b\; \mu_c\,]^T$.

The equations proposed in [7] are reviewed below based on the typical waveforms under ZCS and ZVS conditions shown in Fig. 5.3. These equations include the dead time, T_d, which is from t_3 to t_4 in Fig. 5.3, the switching frequency, f_s, the Root Mean Square (RMS) value of the resonant current, I_{r_rms}, the peak magnetizing current, I_{mpk}, and the corresponding peak flux density, B_{pk}, the RMS value of the current through the parasitic capacitance of the transformer, I_{cp_rms}, and the output current, I_o. The total loss will then be calculated using these equations.

The voltage gain of *LCLC* resonant converter can be found in [7].

$$V_o/V_{in} = 2/a \tag{5.1}$$

The dead time of the *LCLC* resonant converter, T_d, can be calculated by

$$T_d = 1/(2\pi f_{rp})\{\arccos[-\left(\sqrt{(\pi f_{rp})/(2 f_{rs})}\right)^{-1}] - \arctan[(\pi f_{rp})/(2 f_{rs})]\} \tag{5.2}$$

where f_{rs} is the resonant frequency between L_r and C_s, f_{rp} is the resonant frequency between L_m and $C_{p.}$ The switching frequency of the *LCLC* resonant converter can be calculated by

$$f_s = \frac{1}{2}\{\pi\sqrt{L_r C_s} + \sqrt{L_m C_p} \times \{\arccos[-(\sqrt{1 + (\pi^2 L_r C_s)/(4 L_m C_p)})^{-1}] - \arctan(\pi/2\sqrt{(L_r C_s)/(L_m C_p)})\}\,\} \tag{5.3}$$

The RMS value of the resonant current is

$$I_{r\_rms} = (\pi V_o)/(a R_o)\sqrt{f_{rs}/(2 f_s)} \tag{5.4}$$

where R_o is the load of the converter, a is the turns ratio of the transformer.

The peak magnetizing current can be found from

$$I_{mpk} = V_{in}/(2L_m)\sqrt{1/(4 f_{rs}^2) + 1/(\pi^2 f_{rp}^2)} \tag{5.5}$$

The core loss of the transformer can be computed using the peak magnetic flux density, B_{pk}, which can be calculated by

$$B_{pk} = V_{in}\sqrt{1/(4 f_{rs}^2) + 1/(\pi^2 f_{rp}^2)}/(2 N_p A_e) \tag{5.6}$$

where N_p is the number of the turns in the primary winding, A_e is the cross-sectional area of the magnetic core.

The RMS current of the transformer's parasitic capacitance, I_{cp_rms}, which will be utilized to determine the transformer's dielectric loss, is provided by

$$I_{cp\_rms} = V_{in}/L_m\sqrt{0.5f_s/(4f_{rs}^2 + \pi^2 f_{rp}^2)\int_0^{T_d} \sin^2[2\pi f_{rp}(t) + \varphi_m]dt} \quad (5.7)$$

where tan $\varphi_{\mathrm{m}} = \pi f_{rp}/2f_{rs}$.

The output current can be computed by

$$I_o = V_o/R_o \quad (5.8)$$

The power loss in each component will be computed in the section that follows. The total power loss of the *LCLC* resonant converter will then be determined after that.

5.2.2 Power Loss Analysis of the LCLC Resonant Converter in the Space TWTA Applications

As previously discussed, even though ZVS and ZCS of the main switches and rectifiers are attained, the total power loss (P_{tot}) in the *LCLC* resonant converter still includes the conduction losses of the rectifiers (P_D), the driving losses of the main switches (P_{s_dr}), the conduction losses of the main switches (P_{s_on}), the copper losses of the transformers (P_{T_Cu}), the core losses of the transformers (P_{T_Fe}) and the dielectric loss of the transformer (P_{T_Die}). The power loss in each component and the total power loss will be determined in this section based on the review in the previous section.

(a) *Power loss of the high-voltage rectifiers* (P_D)

When Silicon Carbide (SiC) diodes are used as the rectifiers in the *LCLC* resonant converter, the reverse recovery time of SiC diodes can be omitted, therefore the power loss of the rectifiers is the conduction loss, which can be computed by

$$P_D = 2I_o V_D \quad (5.9)$$

where V_D is the forward voltage of the SiC diode.

Combing (5.8) and (5.9), the power loss of the rectifiers can be found from

$$P_D = 2V_D V_o/R_o \quad (5.10)$$

(b) *Driving loss of the main switches* (P_{s_dr})

The driving loss can be computed by

$$P_{s\_dr} = 4Q_g V_{gs} f_s \tag{5.11}$$

where V_{gs} is the driving voltage, Q_g is the gate charge of the MOSFETs.

(c) *Conduction loss of the main switches* (P_{s_on})

The conduction loss can be computed by

$$P_{s\_on} = 2I_{r\_rms}^2 R_{s\_on} \tag{5.12}$$

where R_{s_on} is the on-resistance of the main switch.

Combing (5.4) and (5.12), the conduction loss can be calculated by

$$P_{s\_on} = 2I_{r\_rms}^2 R_{s\_on} \tag{5.13}$$

(d) *Copper loss of the transformer* (P_{T_Cu})

The copper loss of the transformer is

$$P_{T\_Cu} = I_{r\_rms}^2 R_{ac} \tag{5.14}$$

where R_{ac} is the AC resistance of the transformer referred to the primary side.

Combing (5.4) and (5.14), the copper loss of the transformer can be computed by

$$P_{T\_Cu} = (\pi^2 V_o^2 f_{rs} R_{ac})/(2a^2 R_o^2 f_s) \tag{5.15}$$

(e) *Core loss of the transformer* (P_{T_Fe})

The core loss can be calculated based on the Steinmetz equation, which is

$$P_{T\_Fe} = k_c f_s^\alpha (B_{pk})^\beta V_e \tag{5.16}$$

where k_c, α, β are the parameters of the magnetic material, V_e is the volume of the magnetic core.

Combing (5.6) and (5.16), the core loss is

$$P_{T\_Fe} = k_c f_s^\alpha [V_{in}/(2N_p A_e)\sqrt{1/(4f_{rs}^2) + 1/(\pi^2 f_{rp}^2)}]^\beta V_e \tag{5.17}$$

(f) *Dielectric loss of the transformer* (P_{T_Die})

Large amounts of electric energy are stored in the insulator in high-voltage applications. Electric energy is charged and discharged, which results in a dielectric loss. The corresponding parasitic capacitance, or C_p in the *LCLC* resonant converter, is referred to as the primary side. By using the specification of the power loss factor (tanδ), it is possible to compute the Equivalent Series Resistor (R_s) of C_p:

$$\tan\delta = 2\pi f_s C_p R_s \tag{5.18}$$

As a result, R_s can be computed by

$$R_s = \tan\delta/(2\pi f_s C_p) \tag{5.19}$$

The dielectric loss can be computed by

$$P_{T\_Die} = I^2_{cp\_rms} R_s \tag{5.20}$$

Combing (5.7) and (5.20), the dielectric loss is

$$P_{T\_Die} = \frac{4\pi^3 \tan\delta C_p V_{in}^2 f_{rp}^4}{4f_{rs}^2 + \pi^2 f_{rp}^2} \int_0^{T_d} \sin^2[2\pi f_{rp}(t) + \phi]\mathrm{d}t \tag{5.21}$$

where δ is the power loss angle of the dielectric material, and ϕ is constant.

Based on the above analysis, combing (5.10), (5.11), (5.13), (5.15), (5.17), and (5.21), the total power loss of an *LCLC* resonant converter can be computed by

$$\begin{aligned} P_{tot} = {} & \frac{2V_D V_o}{R_o} + 4Q_g V_{gs} f_s + \frac{\pi^2 V_o^2 f_{rs} R_{s\_on}}{a^2 R_o^2 f_s} + k_c f_s^{\alpha} \left(\frac{V_{in}}{2N_p A_e}\sqrt{\frac{1}{4f_{rs}^2} + \frac{1}{\pi^2 f_{rp}^2}}\right)^{\beta} V_e \\ & + \frac{\pi^2 V_o^2 f_{rs} R_{ac}}{2a^2 R_o^2 f_s} + \frac{4\pi^3 \tan\delta C_p V_{in}^2 f_{rp}^4}{4f_{rs}^2 + \pi^2 f_{rp}^2} \int_0^{T_d} \sin^2[2\pi f_{rp}(t) + \phi]\mathrm{d}t \end{aligned} \tag{5.22}$$

The total power loss of the *LCLC* resonant converter has been determined, as shown by (5.22). In the section that follows, an efficiency-oriented two-stage optimal design methodology of the *LCLC* resonant converter will be presented with the goal of minimizing the total power loss.

5.3 The Proposed Efficiency-Oriented Two-Stage Optimal Design Method of the LCLC Resonant Converter in the Space TWTA Applications

The *LCLC* resonant converter's numerous variables and their mutual couplings make it difficult to optimize the total power loss. In this section, an efficiency-oriented two-stage optimal design methodology is presented as a solution to this issue. The proposed two-stage optimal design method's flowchart is shown in the first section. It consists of two stages: in the first, a GA + PSO algorithm is proposed to reduce total power loss, and the optimal parameters are derived; in the second, the proposed single-layer partially-interleaved structure is used to realize the optimal parameters. In the second section, the first stage is further developed, and an example is provided to assess the efficiency of the proposed GA + PSO algorithm. The second stage is expounded in the third part.

5.3.1 The Proposed Efficiency-Oriented Two-Stage Optimal Design Methodology

The flowchart of the presented two-stage optimal design method is depicted in Fig. 5.4, and it includes the following two steps:

Stage 1 (Extraction of the optimal parameters based on the proposed GA + PSO algorithm): A GA + PSO algorithm is presented in the first stage to reduce the total

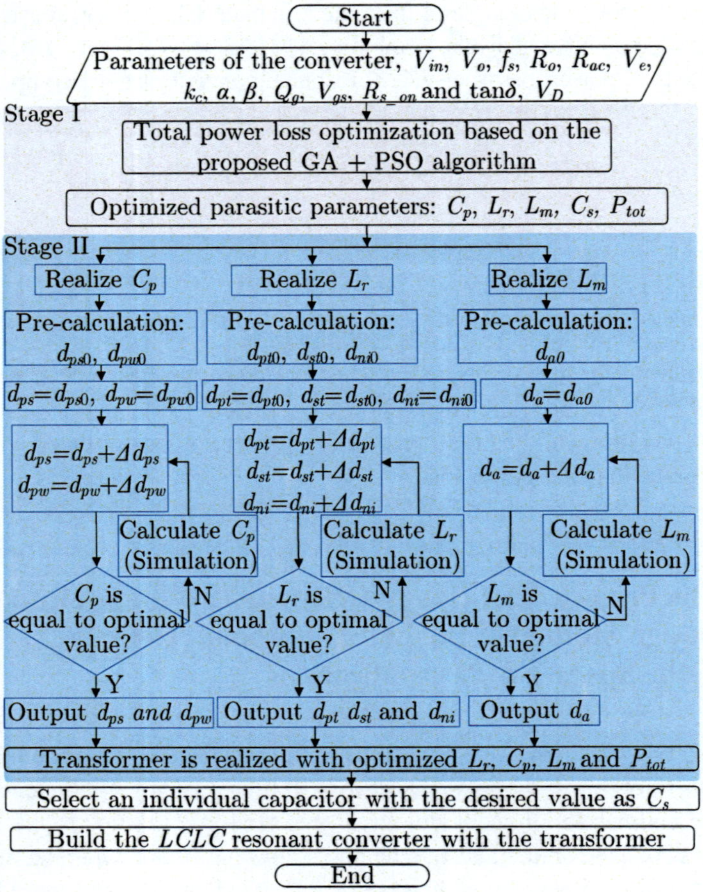

Fig. 5.4 Proposed two-stage optimal design methodology

power loss. The optimum parameters are attained at the end of Stage 1, including the optimum L_r, C_p, L_m, C_s, and P_{tot}, which are utilized to construct the *LCLC* resonant converter.

Stage 2 (Realization of the optimal parameters based on the proposed single-layer partially-interleaved transformer structure): A single-layer partially-interleaved transformer structure is presented since L_r, L_m and C_p are the parasitic parameters of the transformer in order to actualize the ideal parameters. The transformer will be carefully constructed with the presented transformer structure, based on a hybrid electromagnetic analysis, to obtain the optimum L_r, L_m and C_p. The two-stage optimal design approach comes to a conclusion after the transformer design is completed and the entire *LCLC* resonant converter is constructed.

The next sections will go through each of the two stages in Fig. 5.4 in further detail.

5.3.2 Stage-I: Extraction of the Optimal Parameters Based on the Proposed GA + PSO

The presented GA + PSO algorithm is used in the first stage to reduce the total power loss of the *LCLC* resonant converter, and the optimum parameters are then determined.

(a) *Operation principles of the proposed GA + PSO algorithm for the total power loss optimization*

Figure 5.5 depicts the flowchart of the presented GA + PSO algorithm. The summary of the presented GA + PSO algorithm is depicted in Fig. 5.5a. The specifics of the GA part and PSO part are shown in Fig. 5.5b, c, respectively.

The presented GA + PSO algorithm, which is seen in Fig. 5.5a, can be thought of as an enhanced PSO in which the GA optimizes the PSO parameters. The GA is the starting point of the GA + PSO algorithm. First of all, the genes of each individual in GA, which are also the parameters of PSO (*Weight.start*, *Weight.end, kind* and *Vel.max*), are sent to the PSO as the algorithm parameters. The PSO algorithm is then executed given *Weight.start*, *Weight.end*, *kind* and *Vel.max* from GA to extract the number of iterations *N* and total power loss P_{tot}, which is then returned to the GA algorithm. The fitness value of each person in GA is evaluated using *N* and P_{tot}, and the parameters of PSO, *Weight.start*, *Weight.end*, *kind* and *Vel.max* are optimized and passed back to PSO. The recycling process between the GA and PSO will end when the fitness levels in the GA are constant. Following the completion of the recycling process, the optimum *Weight.start*, *Weight.end*, *kind* and *Vel.max* will be sent to the PSO algorithm, which will compute the optimal L_r, C_s, L_m, C_p and P_{tot}, which are also the outputs of the presented GA + PSO method.

The proposed GA + PSO accomplishes the sequential tasks listed below:

Fig. 5.5 Proposed two-stage optimal design methodology

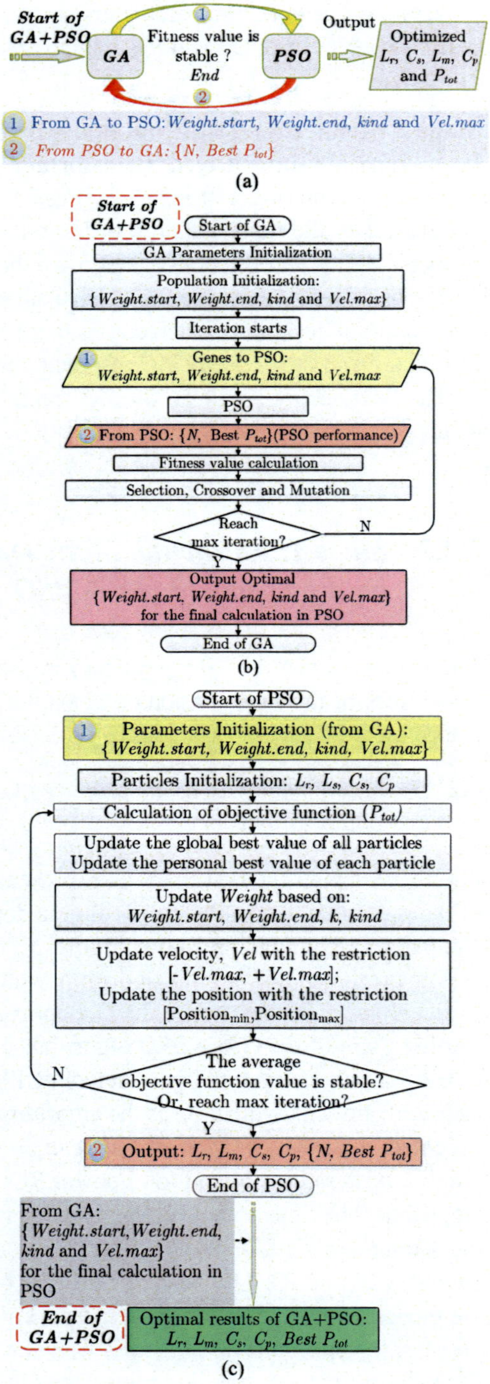

Step 1: Initialization of GA

The parameters of GA are initialized, which include the number of individuals, the crossover rate, the starting mutation rate, the ending mutation rate, and the maximum iteration number. The GA optimizer encodes the parameters of PSO, *Weight.start*, *Weight.end*, *kind* and *Vel.max*, as genes, creates a string of genes to form a chromosome and initializes a starting population.

Step 2: Extraction of the number of iterations and best P_{tot} by invoking the PSO algorithm

Weight.start, *Weight.end*, *kind* and *Vel.max*, are transferred from GA to the PSO algorithm. With the parameters from each individual in GA, the PSO algorithm is performed to extract N and P_{tot}, which will be utilized to evaluate the fitness value of the individual in the GA algorithm. Figure 5.5c contains information about PSO in more detail.

Step 3: Evaluation and Evolution of the population in GA

N and P_{tot}, which are obtained from PSO, are utilized in GA to assess each individual's fitness value. The population in GA evolves through selection, crossover, and mutation based on the fitness value derived from PSO.

Step 4: Repetition of Step 2 and Step 3 to obtain the optimum *Weight.start*, *Weight.end*, *kind* and *Vel.max*

Step 2 to Step 3 are repeated until the maximum iteration number of GA is reached or the average objective function value is stable.

Step 5: Calculations of the optimal design parameters for the *LCLC* resonant converter by PSO algorithm

With the optimum *Weight.start*, *Weight.end*, *kind* and *Vel.max*, the *LCLC* resonant converter is optimized and ideal L_r, C_s, L_m, C_p, and P_{tot} are derived.

The first stage of the presented two-stage ideal design method, which is based on the presented GA + PSO algorithm, is finished when all of the optimum parameters are obtained. The presented GA + PSO algorithm will be assessed with an example in the section that follows.

(b) *Evaluation of the proposed GA + PSO algorithm*

An example is provided in this section to assess the presented GA + PSO algorithm. Table 5.1 lists the parameters for the *LCLC* resonant converter, which has an input voltage of 40 V, an output voltage of 4800 V, a switching frequency of 320 kHz, a rated output power of 288 W, and a rated load of 80 kΩ.

The RJK6505PBF is chosen as the main switch based on the input voltage and the rated output power. Additionally, FEE 38/16/25 with N87 from TDK is chosen as the magnetic core based on the switching frequency and the output power. Moreover, GB01SLT12-214 is chosen as the rectifier after taking the output voltage and the output power into account. In Table 5.1, the parameters of the magnetic core, main switches, and rectifiers are also mentioned.

The presented GA + PSO algorithm selects the total power loss as its objective function and then calculates the ideal L_r, C_s, L_m, C_p, and P_{tot}.

Table 5.1 The parameters of the *LCLC* resonant converter, magnetic core, and main switches

Parameters	Value	Parameters	Value
V_{in}	40.0 V	V_e	10,200 mm^3
V_o	4800 V	A_e	190 mm^2
f_s	320 kHz	k_c	3.716×10^{-24}
R_o	80 kΩ	α	4.823
P_o	288 W	β	5.521
V_{gs}	10.0 V	R_{on}	4.5 mΩ
Q_g	40 nC	R_{ac}	20.0 mΩ
C_{oss}	660 pF		

The optimal design is performed using ACO, BCO, GA, PSO, and the presented GA + PSO in order to compare them to the ACO, BCO, single GA, and single PSO algorithms.

The fluctuation of the total loss with the number of iterations utilizing GA, PSO, and GA + PSO is depicted in Fig. 5.6. As can be observed, the presented GA + PSO and the other two optimal design methods (GA and PSO) both converge to the identical value of 8.9 W. Nevertheless, the GA, PSO, and proposed GA + PSO iteration numbers are 115, 50, 56, 44, and 10, respectively. In light of this, it can be said that the presented GA + PSO algorithm requires a lot fewer iterations than the other two algorithms.

Figure 5.7 displays the computation time for ACO, BCO, GA, PSO, and GA + PSO. It is clear that the presented GA + PSO algorithm takes substantially less time to calculate (1.22 s) than existing algorithms. It can be concluded that the computation time of the presented GA + PSO is significantly shorter than that of ACO, BCO, GA, or PSO.

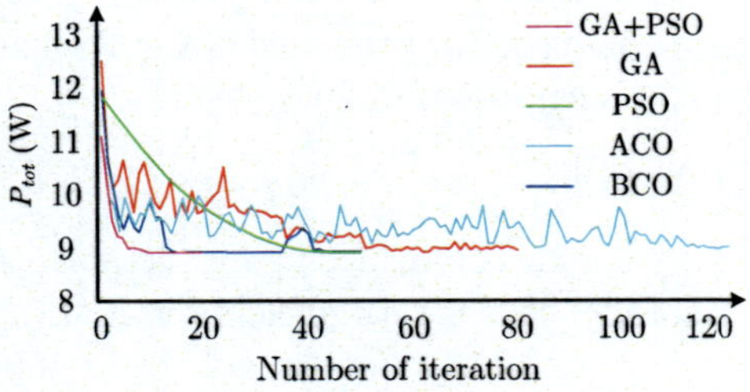

Fig. 5.6 Comparison of iteration number

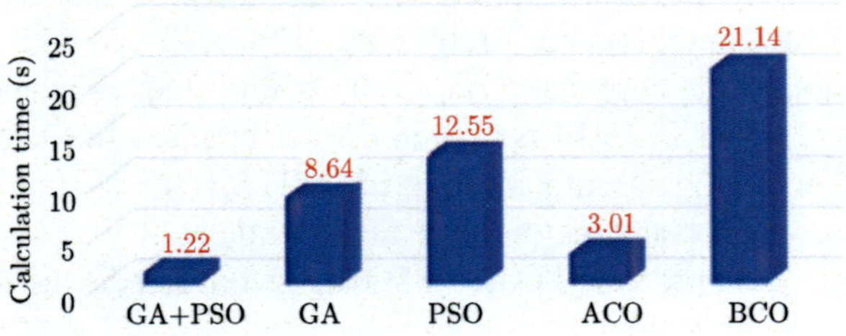

Fig. 5.7 Comparison of calculation time

The presented GA + PSO method performs better than the traditional GA or PSO algorithm, according to the comparison of iteration number and calculation time made above.

The optimum L_r, C_s, L_m, C_p and P_{tot} can be derived based on the ideal results of the presented GA + PSO algorithm, and they are listed as follows: $L_r = 0.09\ \mu$H, $C_s = 1.0\ \mu$F, $L_m = 8.0\ \mu$H, $C_p = 13.2$ nF and $P_{tot} = 8.9$ W.

It should be noticed that C_s, an individual capacitor in the *LCLC* resonant converter, is readily constructed by choosing the suitable capacitance. L_r, L_m and C_p are the transformer's parasitic parameters, nevertheless, and they depend on the structure of the transformer. Consequently, the transformer design should be carried out using the hybrid electromagnetic analysis, which is the second stage of the efficiency-oriented two-stage optimal design method, in order to achieve optimal design results.

5.3.3 Stage-II: Realization of the Optimal Parameters Based on the Proposed Single-Layer Partially-Interleaved Transformer Structure

In this section, a single-layer, partially interleaved transformer construction is presented in order to obtain the optimum design parameters.

In Fig. 5.8, the proposed single-layer partially interleaved transformer structure is depicted, with d_{ps} indicating the distance between primary and the secondary winding; d_{pt} and d_{st} indicate the thickness of the primary and secondary winding; d_{pw} and d_{sw} indicate the width of the primary winding and the secondary winding; d_{ni} and d_{ti} are the thickness of normal insulation and thicker insulation; d_a is the thickness of the air gap. It should be noticed that d_{ti} will be regarded as constant in the transformer design and is designed to satisfy the high-voltage electric insulation requirements in high-voltage applications.

Since the winding configuration [32, 33] determines C_p, L_r and L_m, the parasitic parameters of the transformer, the following procedures will be taken to achieve the ideal values by choosing suitable d_{ps}, d_{pt}, d_{st}, d_{pw} and d_{sw}. It is not necessary to duplicate the detailed design flowcharts of C_p, L_r and L_m provided in the Stage-II of Fig. 5.4.

(a) *Step 1: Design d_{ps}, d_{pw} based on the optimal C_p*

The dielectric layers between the primary winding and the secondary winding are where the majority of the electric energy in the presented transformer configuration is concentrated [32]. Consequently, by choosing the appropriate d_{ps}, d_{pw}, the ideal C_p can be achieved.

The flowchart to calculate d_{ps} and d_{pw} is shown in Fig. 5.4 (Stage II section), and its initial values, d_{ps0} and d_{pw0} are computed based on the analysis of the electric field. The initial parameters, d_{ps0} and d_{pw0}, will then be further changed by electromagnetic

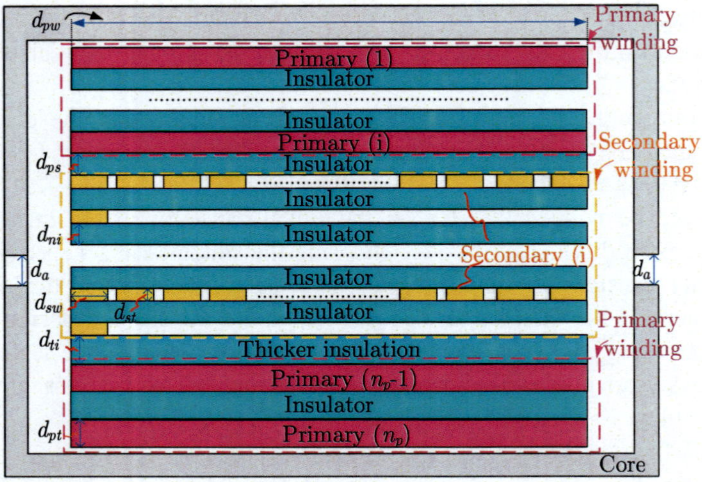

Fig. 5.8 Proposed partially interleaved, single-layer structure

simulations in order to arrive at a more precise C_p. To determine the distribution of the electric field, electrostatic modeling is used. The electric energy, Ee, can be estimated using the distributions of the electric field.

$$E_e = \frac{1}{2} \iiint_V \vec{E} \cdot \vec{D} \mathrm{d}V \tag{5.23}$$

where E and D are the distributions of the electric field.

Additionally, the relationship between the electric energy and the parasitic capacitor, C_p, is

$$E_e = \frac{1}{2} C_p V_p^2 \tag{5.24}$$

where V_p is the voltage across C_p.

Combing (5.23) and (5.24), C_p can be computed based on the distributions of the electric field, which is

$$C_p = \frac{1}{V_p^2} \iiint_V \vec{E} \cdot \vec{D} \mathrm{d}V \tag{5.25}$$

With (5.25), C_p is computed with the distribution of the electric field in Ansys Maxwell. If the optimal C_p is achieved, the values of d_{ps}, d_{pw} will be confirmed; otherwise, the values of d_{ps}, d_{pw} will be updated with the fixed intervals Δd_{ps}, Δd_{pw}.

(b) *Step 2: Design d_{pt}, d_{st}, d_{ni} based on optimal L_r*

By choosing the appropriate d_{pt}, d_{st} and d_{ni}, the ideal leakage inductance can be achieved based on the analysis of the Magneto-Motive Force (MMF) [33]. It should be noticed that the effects of d_{pt}, d_{st}, d_{ni} can be disregarded because the majority of the electric energy is stored in the dielectric layers between the primary winding and the secondary winding.

The flowchart to calculate d_{pt}, d_{st} and d_{ni} is shown in Fig. 5.4 (Stage II section), and its initial values, d_{pt0}, d_{st0} and d_{ni0}, are determined by the analysis of the MMF. The initial values of d_{pt0}, d_{st0} and d_{ni0} will then be further changed by electromagnetic simulations in order to obtain a more precise L_r. To determine the magnetic field distribution, the eddy current simulation is used. Both the primary winding and the secondary winding are subjected to current excitations in order to compute the leakage inductance. The magnetic field energy, E_m, can be computed based on the magnetic field distributions by

$$E_m = \frac{1}{2} \iiint_V \vec{H} \cdot \vec{B} \mathrm{d}V \tag{5.26}$$

where H and B are the distributions of the electric field. Additionally, the relationship between the magnetic field energy and the leakage inductance, L_r, is

$$L_r = \frac{1}{2} L_r i_p^2 \tag{5.27}$$

where i_p is the current applied to the primary winding. Combing (5.26) and (5.27), L_r can be computed based on the distributions of the electric field, which is

$$L_r = \frac{1}{i_p^2} \iiint_V \vec{H} \cdot \vec{B} \mathrm{d}V \tag{5.28}$$

With (5.28), L_r is computed with the distribution of the electric field in Ansys Maxwell. If the optimal L_r is achieved, the values of known d_{pt}, d_{st} and d_{ni} will be confirmed; otherwise, the values of d_{pt}, d_{st} and d_{ni} will be updated with the fixed intervals Δd_{pt}, Δd_{st} and Δd_{ni}.

(c) *Step 3: Design d_a based on optimal L_m*

By adjusting the proposer air gap, d_a, the optimal L_m can be achieved according to the magnetic circuit analysis [33]. Since the variation of d_a has little impact on C_p and L_r in this situation, L_m is the last parameter to be designed.

The flowchart to choose d_a, whose initial values are computed based on the magnetic circuit analysis, is shown in Fig. 5.4 (Stage II section). Further adjustments will be made based on the electromagnetic simulations to minimize the mistakes induced by the calculation. The eddy current simulations are used to determine the magnetic field distribution. The main distinction between L_r and L_m's calculations is that L_m only applies current excitation to the primary winding. Both calculations are based on the magnetic field of the distribution. L_m can be computed based on the distribution of the magnetic field.

$$L_m = \frac{1}{i_p'^2} \iiint_V \vec{H} \cdot \vec{B} \mathrm{d}V \tag{5.29}$$

With (5.29), L_m is computed with the distribution of the magnetic field in Ansys Maxwell. If the optimal L_m is achieved, the value of d_a will be confirmed; otherwise, the d_a value will be updated with the fixed interval Δd_a.

The dimensions of the planar transformer, which are shown in Table 5.1, can be identified using the presented transformer construction and the flowchart in Fig. 5.4.

The planar transformer model is constructed in Ansys Maxwell utilizing parameters provided in Table 5.2, and hybrid electromagnetic simulations are run to verify the parasitic parameters of the planar transformer.

The electric field intensity, which will be applied to compute the parasitic capacitance, and the magnetic field intensity, which will be applied to compute the leakage inductance and the magnetizing inductance, are each depicted in Figs. 5.9 and 5.10, respectively.

The simulated transformer parasitics are computed using the hybrid electromagnetic simulation results presented in Figs. 5.9 and 5.10. Table 5.3 shows the parasitics of simulated transformers. Table 5.3 also includes the optimum parameters acquired from the presented GA + PSO algorithm for comparison purposes.

The optimal parameters of the *LCLC* resonant converter are realized, as shown in Table 5.3, where the simulated parasitics of the presented transformer structure,

Table 5.2 Comparison between the simulated results and the optimal parameters

Parameters	Values	Parameters	Values
d_{ps}	0.13 mm	d_{pt}	0.2 mm
d_{st}	70 μm	d_a	63 μm
d_{pw}	9.6 mm	d_{sw}	0.28 mm
d_{ti}	1.6 mm	d_{ni}	0.3 mm

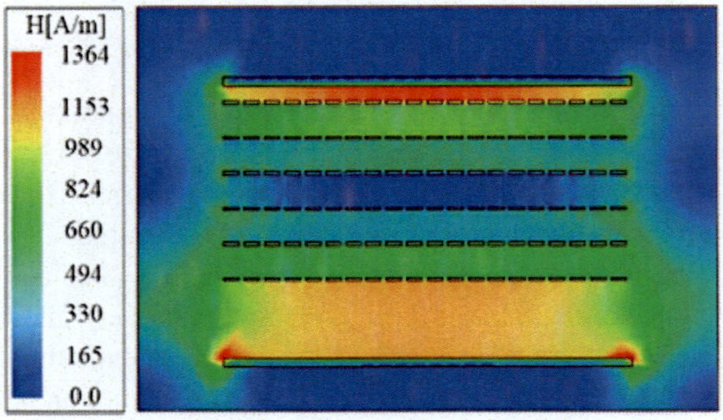

Fig. 5.9 Magnetic field intensity (320 kHz) for the calculations of the leakage inductance (L_r) and the magnetizing inductance (L_m)

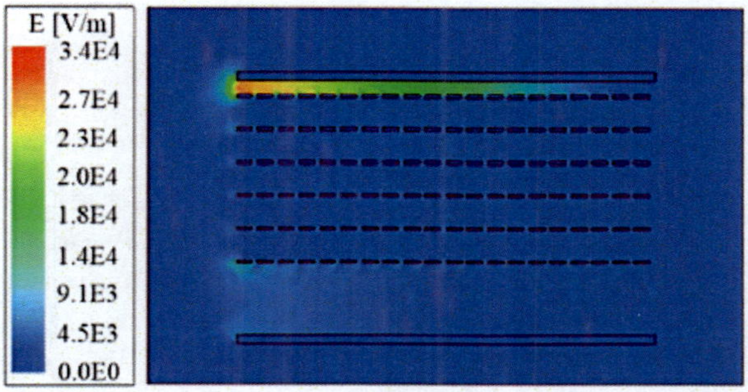

Fig. 5.10 Electric field intensity for the calculation of the parasitic capacitance

Table 5.3 Comparison between the simulated results and the optimal parameters

Parameters	Optimal values	Simulated values
L_r (μH)	0.09	0.09
L_m (μH)	8.0	8.0
C_p (nF)	13.2	13.0

including L_r, L_m, and C_p, are in accordance with the optimum parameters determined by the presented GA + PSO algorithm.

The planar transformer is constructed using the dimensions listed in Table 5.2. Figure 5.11a depicts the three-dimensional (3D) model of the planar transformer. It should be noticed that C_s is an independent component that can be realized by choosing the capacitor with the required value, as opposed to L_r, L_m and C_p, which are the parasitic parameters of the transformer. The ideal C_s in this chapter is a capacitor with a value of 1.0 μF. The presented two-stage optimal design method is used to construct the *LCLC* resonant converter, which is illustrated in Fig. 5.11b, using the planar transformer and the chosen capacitor with the desired value.

Table 5.4 provides a summary of the several component types used in the *LCLC* resonant converter depicted in Fig. 5.11b.

The presented two-stage optimal design approach and the transformer construction will both be put to the test through experiments in the section that follows.

5.4 Experimental Validations

In this part, the presented efficiency-oriented two-stage optimal design method is tested with the *LCLC* resonant converter shown in Fig. 5.11b. Figure 5.12 depicts the complete circuit of the two-stage power converter for the space TWTA applications. A boost converter serves as the first stage, and an *LCLC* resonant converter serves as the second. The first stage of a two-stage power converter controls the output voltage while the second stage functions as a DC transformer in an open-loop configuration. Additionally, the two-stage power converter's output voltage is controlled by a PI controller. The bus voltage, V_{bus}, ranges from 25.2 to 30.8 V in this chapter. The

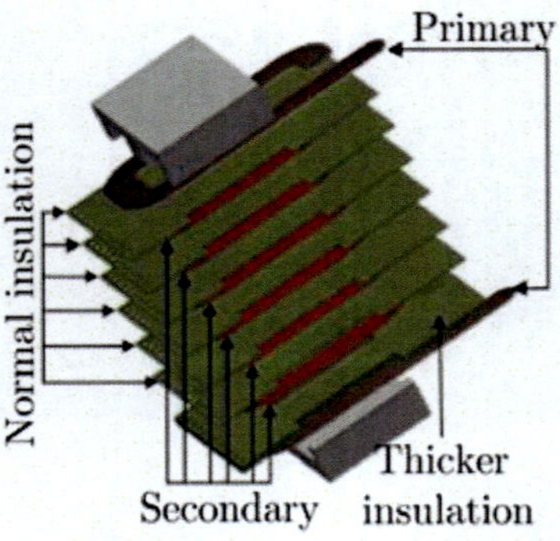

(a)

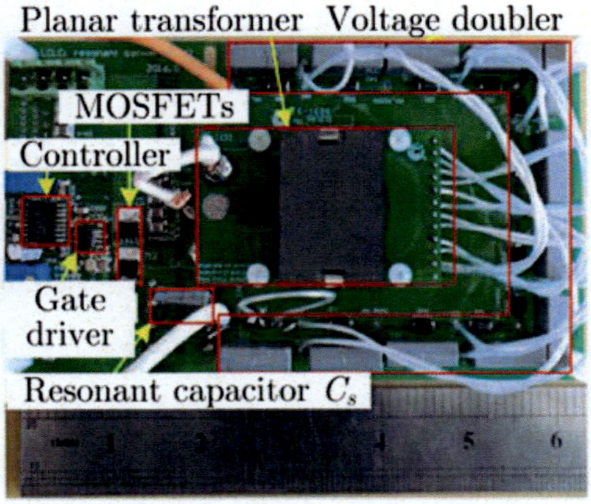

(b)

Fig. 5.11 The built experimental platform: (**a**) The 3-dimensonal (3D) model of the planar transformer, (**b**) the LCLC resonant converter

Table 5.4 Types of the components in the *LCLC* resonant converter

Component	Type	Component	Type
S_1, S_2, S_3, S_4	RJK6505PBF	C_s	Polypropylene
D_1, D_2	GB01SLT12-214	C_1, C_2	Polypropylene
Gate driver	UCC 27210	Core	FEE38/16/25(N87)

LCLC resonant converter's input voltage, V_{in}, as well as the Boost converter's output voltage, are both controlled to 40 V. Then, using the *LCLC* resonant converter, V_{in} will be increased to the high voltage output, V_o, for the TWTA space, which is 4800 V.

The validation of the soft switching of the main switches and rectifiers occurs in the first section. The efficiency of the presented two-stage optimal design method is assessed in the second section by testing the scenarios when C_s and L_m stray from their corresponding optimal values. The effectiveness of the optimized *LCLC* resonant converter is assessed in the third section under various input voltages and loads.

5.4.1 Verifications of the ZVS and ZCS Characteristics of the Optimal LCLC Resonant Converter

The *LCLC* resonant converter was tested at its rated power of 288 W. Figure 5.13 depicts the experimental waveforms of the first step. The input voltage of the first stage, V_{bus}, is 28 V, and the output voltage of the first stage, which is also the input of the second stage, V_{in}, is 40 V, as shown in Fig. 5.13.

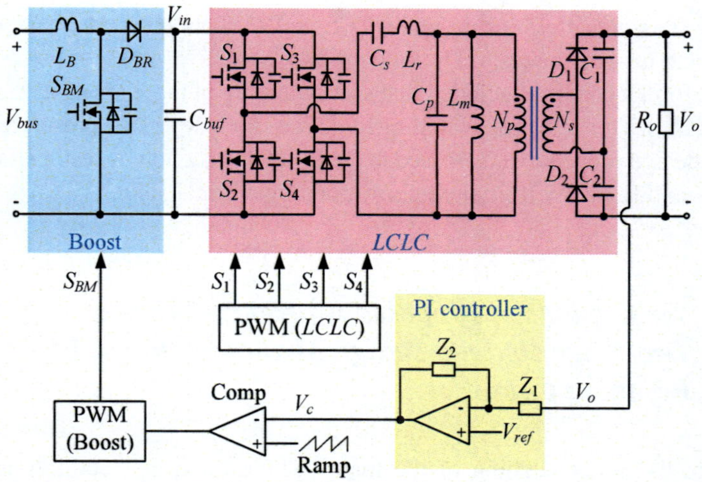

Fig. 5.12 The circuit of the two-stage power converter for the space TWTA applications

Fig. 5.13 Waveforms of the first stage

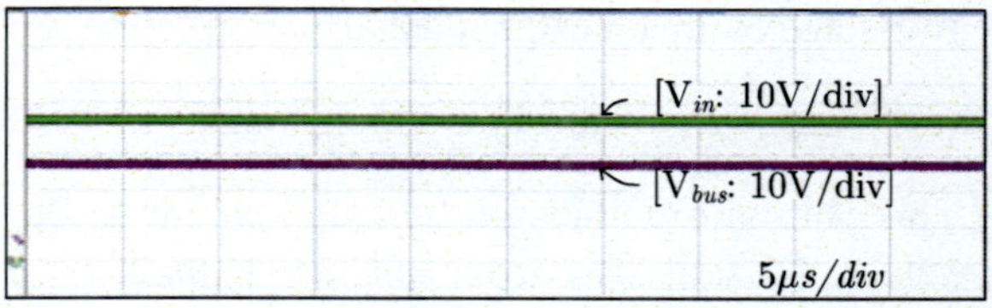

Figure 5.14a depicts the driving signals of S_1, $v_{gs1}(t)$, the resonant current, $i_r(t)$, and the voltage of C_{ds1}, $v_{ds1}(t)$. Figure 5.14b depicts the current of D_1, $i_{d1}(t)$, the voltage of D_1, $v_{d1}(t)$. As shown in Fig. 5.14a, when S_1 is switched on, $v_{ds1}(t)$ and $i_r(t)$ are both 0. Therefore, when S_1 is switched on, it runs under ZCS and ZVS circumstances. Furthermore, when S1 is switched off, $v_{ds1}(t)$ and $i_r(t)$ are both 0. Consequently, when S_1 is switched off, it runs under ZCS and ZVS circumstances. As a result, the turn-on and turn-off losses of S_1 are both decreased.

Similarly, as shown in Fig. 5.14b, when D_1 is switched on, $v_{d1}(t)$ and $i_{d1}(t)$ are both 0. Therefore, when D_1 is switched on, it runs under ZCS and ZVS circumstances. Additionally, when D_1 is switched off, $v_{ds1}(t)$ and $i_r(t)$ are both 0. Consequently,

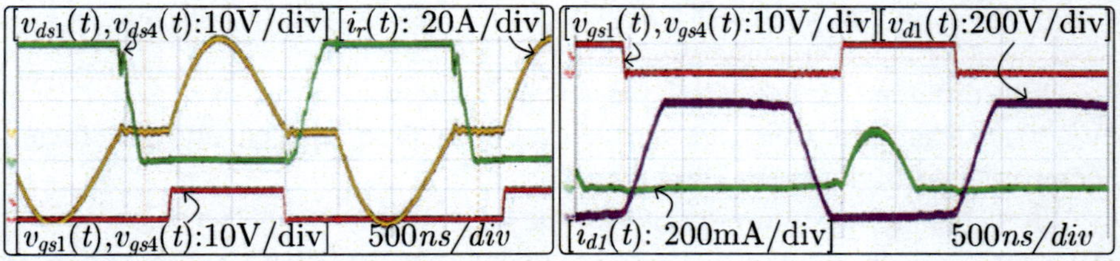

Fig. 5.14 Soft switching experimental waveforms: **a** S_1; **b** D_1

when D_1 is switched off, it runs under ZCS and ZVS circumstances. As a result, the turn-on and turn-off losses of D_1 are both decreased.

Therefore, it can be concluded that soft switching of the main switches and rectifiers results in a reduction of switching loss. The efficiency of the presented two-stage optimal design procedure is assessed in the following section by testing the scenarios when Cs and Lm stray from their corresponding optimal values.

5.4.2 *Verifications of Proposed Efficiency-Oriented Two-Stage Optimal Design Method of the* LCLC *Resonant Converter*

Although the soft switching of the main switches and the rectifiers is achieved, it is still necessary to analyze the effectiveness of the overall power converter. The scenarios where Cs and Lm deviate from their respective optimum value are evaluated in this section in order to validate the efficiency of the presented efficiency-oriented two-stage optimal design method.

(a) *Case I: C_s deviates from the optimal value* (1.0 μF)

This portion examines the situation where C_s deviates from its ideal value. Two values of C_s are chosen, one of which is smaller than the ideal design result (0.3 μF) and the other bigger than the ideal design result (1.6 μF).

With different C_s (0.3 μF and 1.6 μF), the waveforms of $v_{gs1}(t)$, $i_r(t)$ and $v_{ds1}(t)$ are examined, which are depicted in Fig. 5.15a, b respectively. The efficiency comparison is displayed in Fig. 5.15c.

From Fig. 5.15a, b, it is clear that the advantages of soft switching may lose because of the deviation of C_s from the optimal value (No ZVS in Fig. 5.15a and no ZCS in Fig. 5.15b). Additionally, based on the efficiency comparison in Fig. 5.15c, the efficiency with the optimal C_s is the highest compared to the scenarios where C_s deviates from the optimal design result, validating the efficiency of the presented two-stage optimal design method.

(b) *Case I: Case I: L_m deviates from the optimal value* (8.0 μH)

This portion examines the situation where L_m deviates from its ideal value. Two values of L_m are chosen, one of which is smaller than the ideal design result (4.0 μH) and the other bigger than the ideal design result (29.3 μH).

With different L_m (29.3 μH and 4.0 μH), the waveforms of $v_{gs1}(t)$, $i_r(t)$ and $v_{ds1}(t)$ are examined, which are depicted in Fig. 5.16a, b, respectively. The efficiency comparison is displayed in Fig. 5.16c.

From Fig. 5.16a, b, it is clear that the advantages of soft switching may lose because of the deviation of L_m from the optimum value (No ZVS in Fig. 5.16a and no ZCS in Fig. 5.16b). Additionally, based on the efficiency comparison in Fig. 5.16c, the efficiency with the optimal L_m is the highest compared to the scenarios where

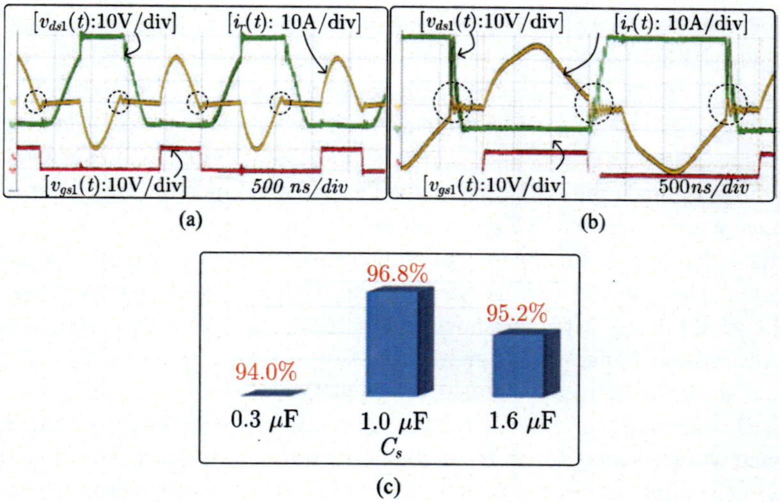

Fig. 5.15 Experimental results with different C_s: **a** $C_s = 0.3$ μF, **b** $C_s = 1.6$ μF, **c** efficiency comparison

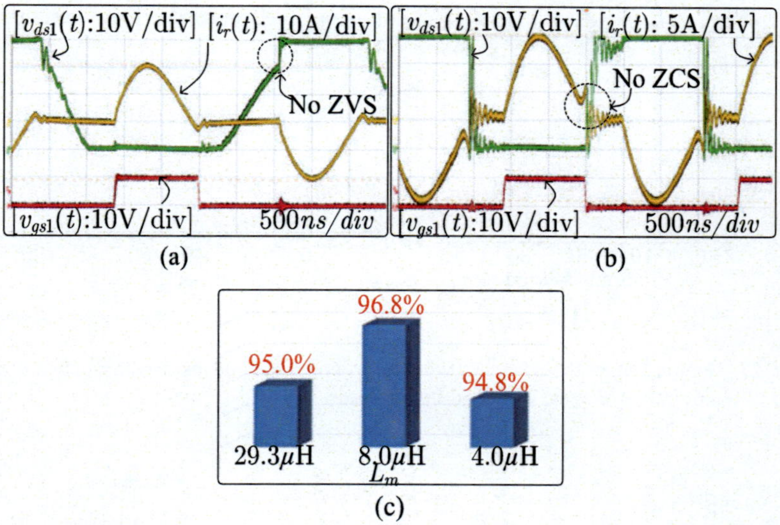

Fig. 5.16 Experimental results with different L_m: **a** $L_m = 29.3$ μH, **b** $L_m = 4.0$ μH, **c** comparison of the efficiency

L_m deviates from the optimal design result, validating the efficiency of the presented two-stage optimal design method.

(c) *Efficiency under different input voltages and loads*

The efficiency of the *LCLC* resonant converter with the presented two-stage optimal design method and the conventional design method under various input voltages is displayed in Fig. 5.17a [7]. The total measured power loss at the rated input voltage (40 V) based on the presented two-stage optimal design method is 9.22 W and the efficiency is 96.8%. The measured total power loss (9.22 W) is quite close to the 8.9 W optimal outcome of the two-stage optimal design method based on GA + PSO that is presented. Additionally, it can be deduced that the *LCLC* resonant converter with the presented optimal design method has greater efficiency under various input voltages when compared to the conventional design method.

The efficiency of the *LCLC* resonant converter with the presented two-stage optimal design method and the conventional design method under various loads is displayed in Fig. 5.17b [7]. As shown in Fig. 5.17b that the *LCLC* resonant converter with the presented optimal design method has greater efficiency under various loads when compared to the conventional design method.

The loss breakdown of the *LCLC* resonant converter with the proposed efficiency-oriented two-stage optimal design method and the conventional design method under rated input voltage and power is displayed in Fig. 5.18. It is evident that the conduction loss, copper loss, core loss, and dielectric loss of the conventional design method are higher than those of the proposed optimal design method.

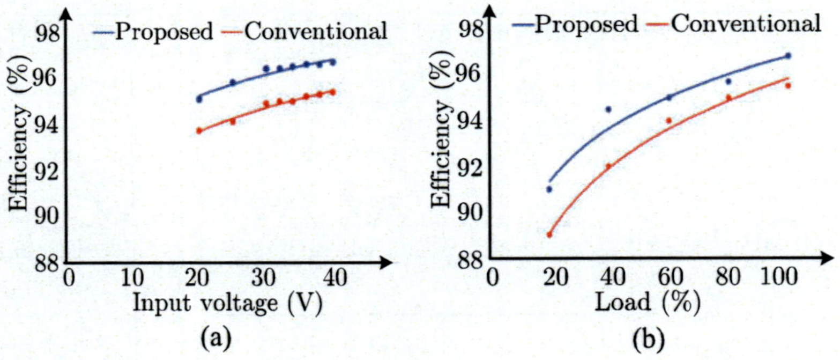

Fig. 5.17 Comparison of the efficiency with the proposed method and the conventional method: **a** under different input voltages; **b** under different loads

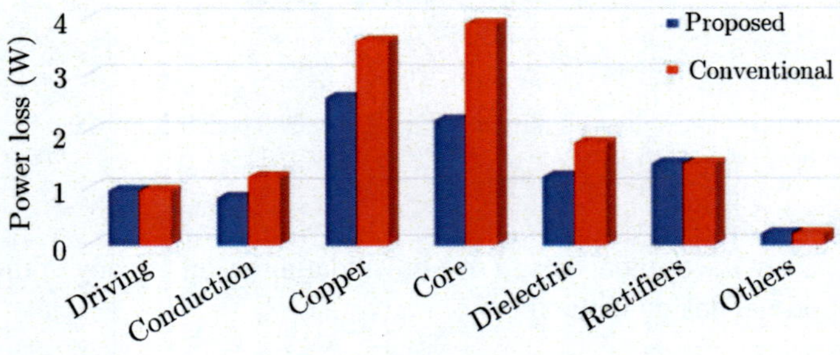

Fig. 5.18 Loss breakdown of the *LCLC* resonant converter under rated power with the proposed method and the conventional method

5.5 Conclusions

An efficiency-oriented two-stage optimal design methodology of high-efficiency *LCLC* resonant converters for space TWTA applications have been presented. The *LCLC* resonant converter and the corresponding magnetic design are intended to have their total power loss optimized using the presented optimal design method.

The presented GA + PSO algorithm is used in the first stage to optimize the total power loss of the *LCLC* resonant converter, which includes the driving loss, conduction loss of the switches, copper loss, core loss, and the dielectric loss of the transformer, as well as the conduction loss of the rectifiers. From there, the optimal L_r, L_{m}, C_s, C_p, and P_{tot} are obtained.

The planar transformer with the ideal parameters is designed in the second stage in order to achieve the optimal parameters discovered in the first stage. The design flowchart for the proposed single-layer, partially-interleaved transformer structure is also provided. The transformer structure that has been presented can provide the optimum design results. The presented transformer structure was supported by both the results of the experiments and the hybrid electromagnetic simulations.

Finally, the presented two-stage optimal design method of the *LCLC* resonant converter has been validated by the experiments.

References

1. E. F. Nicol and J. M. Robison, "TWTA On-Orbit Reliability for Satellite Industry," *IEEE Trans. on Electron Devices*, vol. PP, no. 99, pp. 1–5.
2. A. Arora, M. Thottappan and P. K. Jain, "Design and Stability Studies of Second-Harmonic Gyro-TWT Amplifier Using Wedge-Shaped Lossy Ceramic Rod-Loaded Mode Selective RF Interaction Circuit," *IEEE Trans. on Plasma Science*, vol. 44, no. 10, pp. 2340–2347, Oct. 2016.
3. N. V. Bijeev, Anju Malhotra, Virender Kumar, Surinder Singh, K. S. Dasgupta, Rajesh N. Motta, B. Venugopal, Sandhyarani, O. K. Jinan, B. K. Jayakumar, "Design and realization challenges of power supplies for space TWT," *2011 IEEE International Vacuum Electronics Conference (IVEC)*, Bangalore, 2011, pp. 431–432.
4. D. Yang and H. Yin, "Energy Conversion Efficiency of a Novel Hybrid Solar System for Photovoltaic, Thermoelectric, and Heat Utilization," in *IEEE Trans. on Energy Conversion*, vol. 26, no. 2, pp. 662–670, June 2011.
5. M. Tabib-Azar, O. C. Fawole, S. S. Pandey and C. H. Mastrangelo, "Microplasma Traveling Wave Terahertz Amplifier," *IEEE Trans. on Electron Devices*, vol. 64, no. 9, pp. 3877–3884, Sept. 2017.
6. I. Barbi and R. Gules, "Isolated DC-DC converters with high-output voltage for TWTA telecommunication satellite applications," *IEEE Trans. on Power Electronics*, vol. 18, no. 4, pp. 975–984, July 2003.
7. B. Zhao, G. Wang and W. G. Hurley, "Analysis and Performance of *LCLC* Resonant Converters for High-Voltage High-Frequency Applications," *IEEE Journal of Emerging and Selected Topics in Power Electronics*, vol. 5, no. 3, pp. 1272–1286, Sept. 2017.
8. X. Zhang, Q. Zhong, V. Kadirkamanathan, J. He and J. Huang, "Source-side Series-virtual-impedance Control to Improve the Cascaded System Stability and the Dynamic Performance of Its Source Converter," in IEEE *Trans. on Power Electronics*. 2018. doi: https://doi.org/10.1109/TPEL.2018.2867272.

9. X. Zhang and Q. Zhong, "Improved Adaptive-Series-Virtual-Impedance Control Incorporating Minimum Ripple Point Tracking for Load Converters in DC Systems," in IEEE Transactions on Power Electronics, vol. 31, no. 12, pp. 8088–8095, Dec. 2016.
10. N. Shafiei, M. Pahlevaninezhad, H. Farzanehfard and S. R. Motahari, "Analysis and Implementation of a Fixed-Frequency *LCLC* Resonant Converter With Capacitive Output Filter," *IEEE Trans. on Industrial Electronics*, vol. 58, no. 10, pp. 4773–4782, Oct. 2011.
11. R. L. Lin and L. H. Huang, "Efficiency Improvement on *LLC* Resonant Converter Using Integrated *LCLC* Resonant Transformer," *IEEE Trans. on Industry Applications*, vol. 54, no. 2, pp. 1756–1764, March-April 2018.
12. C. Liu, F. Teng, C. Hu and Z. Zhang, "*LCLC* resonant converter for multiple lamp operation ballast," *Applied Power Electronics Conference and Exposition, 2003. APEC '03.* Eighteenth Annual IEEE, Miami Beach, FL, USA, 2003, pp. 1209–1213 vol.2.
13. C. M. Bingham, Y. A. Ang, M. P. Foster and D. A. Stone, "Analysis and Control of Dual-Output *LCLC* Resonant Converters with Significant Leakage Inductance," *IEEE Trans. on Power Electronics*, vol. 23, no. 4, pp. 1724–1732, July 2008.
14. J. Liu, K. W. E. Cheng, J Zeng, " A Unified Phase-Shift Modulation for Optimized Synchronization of Parallel Resonant Inverters in High Frequency Power System," *IEEE Trans. on Industrial Electronics*, vol. 61, no. 7, pp. 3232–3247, July 2014.
15. Z. Ye, P. K. Jain and P. C. Sen, "A Two-Stage Resonant Inverter with Control of the Phase Angle and Magnitude of the Output Voltage," *IEEE Trans. on Industrial Electronics*, vol. 54, no. 5, pp. 2797–2812, Oct. 2007.
16. W. Chen, Y. Rao, C. Shan, G. Fujita and T. Yasutoshi, "The Design and Experiment of Ion Generator Power Supply for Vacuum Sputtering," 2007 Power Conversion Conference—Nagoya, Nagoya, 2007, pp. 931–935.
17. Y. Wang, J. M. Alonso and X. Ruan, "A Review of LED Drivers and Related Technologies," *IEEE Trans. on Industrial Electronics*, vol. 64, no. 7, pp. 5754–5765, July 2017.
18. W. Guo and P. K. Jain, "A low frequency AC to high frequency AC inverter with build-in power factor correction and soft-switching," *IEEE Trans. Power Electronics*, vol. 19, no. 2, pp. 430–442, Mar. 2004.
19. S. Park and J. Rivas-Davila, "Power loss of GaN transistor reverse diodes in a high frequency high voltage resonant rectifier," *2017 IEEE Applied Power Electronics Conference and Exposition (APEC)*, Tampa, FL, 2017, pp. 1942–1945.
20. J. M. Lopera, M. J. Prieto, J. Díaz and J. García, "A Mathematical Expression to Determine Copper Losses in Switching-Mode Power Supplies Transformers Including Geometry and Frequency Effects," *IEEE Trans. on Power Electronics*, vol. 30, no. 4, pp. 2219–2231, April 2015.
21. J. R. Sibue, G. Meunier, J. P. Ferrieux, J. Roudet and R. Periot, "Modeling and Computation of Losses in Conductors and Magnetic Cores of a Large Air Gap Transformer Dedicated to Contactless Energy Transfer," *IEEE Trans. on Magnetics*, vol. 49, no. 1, pp. 586–590, Jan. 2013.
22. J. Hao, J. Fu, Z. Ma, S. Zhang and S. Shen, "Condition assessment of main insulation in transformer by dielectric loss data interpolation method and database building," *Proceedings of 2014 International Symposium on Electrical Insulating Materials*, Niigata, 2014, pp. 152–155.
23. X. Lai, P. Zhang, Y. Wang and M. Wu, "Position-Posture Control of a Planar Four-Link Underactuated Manipulator Based on Genetic Algorithm," *IEEE Trans. on Industrial Electronics*, vol. 64, no. 6, pp. 4781–4791, June 2017.
24. S. E. De León-Aldaco, H. Calleja and J. Aguayo Alquicira, "Metaheuristic Optimization Methods Applied to Power Converters: A Review," *IEEE Trans. on Power Electronics*, vol. 30, no. 12, pp. 6791–6803, Dec. 2015.
25. S. H. Chung and H. K. Chan, "A Two-Level Genetic Algorithm to Determine Production Frequencies for Economic Lot Scheduling Problem," *IEEE Trans. on Industrial Electronics*, vol. 59, no. 1, pp. 611–619, Jan. 2012.
26. Z. Yin, C. Du, J. Liu, X. Sun and Y. Zhong, "Research on Auto disturbance-Rejection Control of Induction Motors Based on an Ant Colony Optimization Algorithm," *IEEE Trans. on Industrial Electronics*, vol. 65, no. 4, pp. 3077–3094, April 2018.

27. Y. Zhou, "Runtime Analysis of an Ant Colony Optimization Algorithm for TSP Instances," *IEEE Trans. on Evolutionary Computation*, vol. 13, no. 5, pp. 1083–1092, Oct. 2009.
28. A. Karaarslan, "The Implementation of Bee Colony Optimization Algorithm to Sheppard–Taylor PFC Converter," *IEEE Trans. on Industrial Electronics*, vol. 60, no. 9, pp. 3711–3719, Sept. 2013.
29. F. S. Abu-Mouti and M. E. El-Hawary, "Optimal Distributed Generation Allocation and Sizing in Distribution Systems via Artificial Bee Colony Algorithm," *IEEE Trans. on Power Delivery*, vol. 26, no. 4, pp. 2090–2101, Oct. 2011.
30. M. Veerachary and A. R. Saxena, "Optimized Power Stage Design of Low Source Current Ripple Fourth-Order Boost DC–DC Converter: A PSO Approach," *IEEE Trans. on Industrial Electronics*, vol. 62, no. 3, pp. 1491–1502, March 2015.
31. H. Shi, H. Wen, Y. Hu and L. Jiang, "Reactive Power Minimization in Bidirectional DC–DC Converters Using a Unified-Phasor-Based Particle Swarm Optimization," *IEEE Trans. on Power Electronics*, vol. 33, no. 12, pp. 10990–11006, Dec. 2018.
32. B. Zhao, Z. Ouyang, M. C. Duffy, M. A. E. Andersen and W. G. Hurley, "An Improved Partially Interleaved Transformer Structure for High-Voltage High-Frequency Multiple-Output Applications," *IEEE Trans. on Industrial Electronics*, vol. 66, no. 4, pp. 2691–2702, April 2019.
33. Z. Ouyang, O. C. Thomsen and M. A. E. Andersen, "Optimal Design and Tradeoff Analysis of Planar Transformer in High-Power DC–DC Converters," *IEEE Trans. on Industrial Electronics*, vol. 59, no. 7, pp. 2800–2810, July 2012.

Chapter 6
The Proposed AI Based Two-Stage Optimal Design Methodology for High-Efficiency Bidirectional Power Converters in the Hybrid AC/DC Microgrid Applications

6.1 Introduction

The hybrid AC/DC microgrid, which combines the DC sub-microgrid and AC sub-microgrid, is becoming increasingly well-liked as more and more renewable energy sources, such as PV panels and wind turbines, are integrated into the power system [1–3]. The bus conversion system is necessary to connect the AC and DC bus because the AC/DC sources and loads are integrated. The basic design of a hybrid AC/DC microgrid is depicted in Fig. 6.1 [4], which interconnects the AC and DC bus using a bidirectional interlinking converter (BIC) and a DC transformer.

A bidirectional converter is the BIC [5]. BIC functions as an inverter when power is flowing from a DC to an AC bus and as a rectifier when it is flowing from an AC to a DC bus. As for the DC transformer, due to their benefits like high power density, high efficiency, buck and boost capabilities, and bidirectional power transmission, dual active bridge (DAB) converters and symmetrical *CLLC* resonant converters (see Fig. 6.2) are typically used [6, 7]. Nevertheless, the superiority of Zero Voltage Switching (ZVS) for DAB converters will vanish under low load situations. While the symmetrical *CLLC* resonant converter can achieve ZVS for the primary main switches and Zero-Current-Switching (ZCS) for the secondary rectifiers [8].

This chapter attention to the symmetrical *CLLC* resonant converter as the DC transformer in the hybrid AC/DC microgrid applications to accomplish soft switching under all load circumstances.

Reviews of earlier studies on the symmetrical *CLLC* resonant converter's topology [9–12], controls [9, 13], and magnetic design [13–17] can be found here. The complete explanations of the working principles and equations governing the circuit parameters can be found in [9, 10]. Additionally, a design method for the *CLLC* resonant converter based on voltage gain is presented. The *CLLC* resonant converter's robust circuit parameters design with the open-loop control for the optimum power transmission and voltage regulation capability is presented in [11]. The conventional *CLLC* resonant converter is combined with a buck/boost converter in [12], where pulse width modulation allows for high voltage gain. Regarding the control

X. Zhang et al., *Automated Design of Electrical Converters with Advanced AI Algorithms*,
https://doi.org/10.1007/978-981-99-0459-4_6

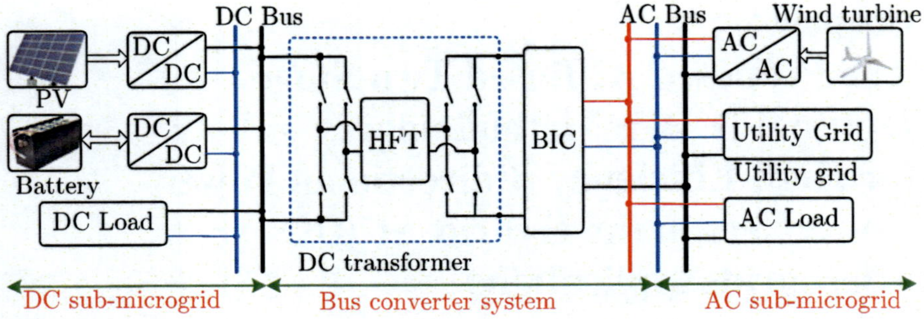

Fig. 6.1 Typical structure of a hybrid AC/DC microgrid

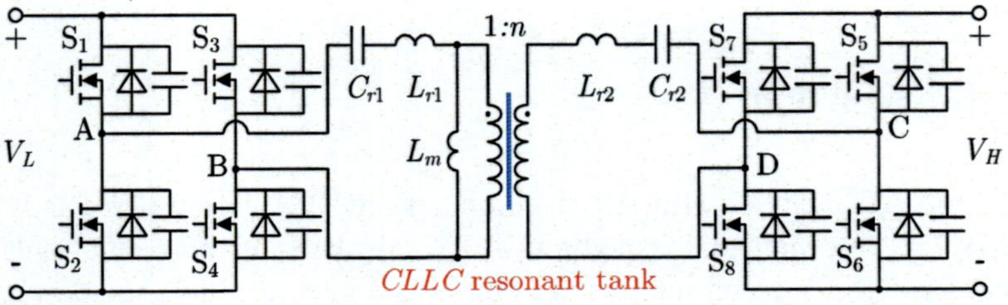

Fig. 6.2 A typical symmetrical *CLLC* resonant converter as a DC transformer

techniques, in [9], an intelligent digital control algorithm is proposed to control bidirectional power conversions and output voltage. In [13], a phase-controlled *CLLC* resonant converter's linear state-space model is developed. This model can be applied for quick simulations and state variable prediction with a high signal variation. A *CLLC* resonant converter has two resonant inductances and a transformer as its magnetic components, as depicted in Fig. 6.2. In addition to the traditional method, which employs two distinct inductances [8], the leakage inductances are utilized as the resonant inductances [14, 15]. In [16], a unique 6-layer PCB winding transformer with integrated resonant inductors is presented for a three-phase *CLLC* resonant converter. Zhao et al. [17] describes the modular power conversion of a high frequency transformer while taking insulation, efficiency, and power density into account.

The primary purpose of the *CLLC* resonant converter in hybrid AC/DC microgrid systems is to transmit power with a high degree of efficiency since the energy management system (EMS) [7] regulates the voltage of the AC and DC bus. Therefore, the conversion efficiency rather than voltage regulation is a difficulty for the *CLLC* resonant converter. The total power loss of the converter requires to be further adjusted in order to reach high efficiency. The switching loss analysis was covered in earlier studies on the *CLLC* resonant converter [12, 17], but the total power loss optimization is the most difficult due to the numerous circuit variables that must be taken into account and their internal couplings. As a result, there are few reports of research on total power loss optimization.

The typical method to increase the power density from the standpoint of the magnetic design in the *CLLC* resonant converter is to employ the leakage inductances as the resonant inductances [14, 15], as separate inductors are not necessary. Nevertheless, the general equations of the distance between the primary winding and secondary winding (d_w), and the thickness of the air gap (d_a) to design the specific leakage inductances and magnetizing inductance, taking insulation and conductor thicknesses into account are not reported.

Due to their propensity for coping with complicated issues and parallelism, AI algorithms are becoming more and more well-liked and frequently utilized in optimal design problems nowadays [18, 19]. In this chapter, the issue brought on by the numerous variables and internal couplings in the *CLLC* resonant converter is resolved using the AI method. The optimization of total power loss and the accompanying magnetic design serve as the objectives of this chapter, which are as follows: This chapter seeks to reduce the total power loss of the symmetrical *CLLC* resonant converter used as a DC transformer using an AI based algorithm, taking into account all switching loss, switch conduction loss, resonant capacitance loss, copper loss, and transformer core loss factors. Prior to that, the calculations of the key parameters and the operating principles of the *CLLC* resonant converter are examined. The total power loss of the *CLLC* resonant converter is then minimized using a two-stage optimal design process based on an AI algorithm: The key parameters are optimized in Stage-1 using a GA (Genetic Algorithm) + PSO (Particle Swarm Optimization) algorithm; in Stage-2, the transformer is designed using a hybrid electromagnetic analysis approach in order to attain the optimum parameters obtained from the AI algorithm. This chapter offers a method for creating the optimum leakage inductances and magnetizing inductances from the perspective of magnetic design. To attain the optimum leakage inductances and magnetizing inductance, the equations of d_w and d_a are derived.

The order of the remaining chapters is as follows. The working theories and calculations of the key parameters of the *CLLC* resonant converter are examined in Sect. 6.2. The bulk of the chapter, Sect. 6.3, focuses on the AI based two-stage optimum design method. At the first, the equation of the total power loss is derived for further optimization; A hybrid AI approach, GA + PSO, is then developed to optimize the total power loss, which is the first step. And then, a planar transformer for the *CLLC* resonant converter is developed using the leakage inductances as the resonant inductances in order to attain the desired ideal parameters of the optimized transformer. The proposed two-stage optimal design method and the d_w and d_a equations are proved by simulations and experiments in Sect. 6.4. The entire chapter is summarized in Sect. 6.5.

6.2 Working Principles and the Circuit Analysis of the CLLC Resonant Converter in Hybrid AC/DC Microgrid Applications

This section analyzes the operating concepts and calculations of the open-loop *CLLC* resonant converter. The *CLLC* resonant converter should run at a 50% duty cycle to obtain high efficiency [20]. Additionally, the switching frequency (f_s) should be a little lower than the resonance frequency (f_r) between C_{r1} and L_{r1} in order to accomplish ZVS and ZCS. Figure 6.3 displays the ZVS and ZCS waveforms of *CLLC* resonant converters.

6.2.1 *Working Principles of the CLLC Resonant Converter as a DC Transformer*

The main waveforms of interest are as follows:

(a) The driving signals of S_1 and S_4, $v_{gs1(4)}$, same as S_5 and S_8; the driving signals of S_2 and S_3, $v_{gs2(3)}$, same as S_6 and S_7, $v_{gs6(7)}$;
(b) the voltage across the parasitic capacitor of S_1, $v_{ds1}(t)$, same as $v_{ds4}(t)$; the voltage across the parasitic capacitor of S_5, $v_{ds5}(t)$, same as $v_{ds8}(t)$;
(c) the current on the primary side, $i_p(t)$; the current on the secondary side, $i_s(t)$;
(d) the magnetizing current, $i_m(t)$;
(e) the voltage across the series capacitor (C_{r1}), $v_{Cr1}(t)$.

The assumptions are made in the analysis as follows:

(a) The input voltage (V_L) and output voltage (V_H) are constant;
(b) All the components are ideal (lossless).

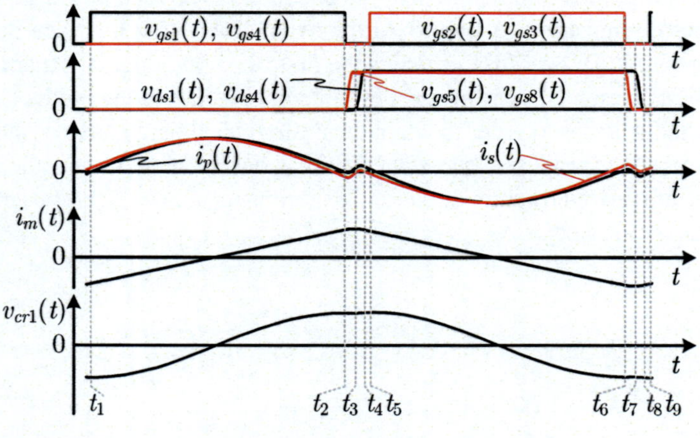

Fig. 6.3 Typical ZVS and ZCS waveforms of a *CLLC* resonant converter

Mode 1 (t_1, t_2): Figure 6.3 displays the waveforms of Mode 1 (from t_1 to t_2), and Fig. 6.4 depicts the equivalent circuit of Mode 1. Prior to Mode 1, the body diodes of S_1 and S_4 conduct the primary current, $i_p(t)$, while S_5 and S_8 conduct the secondary current, $i_s(t)$. Consequently, the voltages of S_1, S_4, S_5, and S_8, as well as $v_{ds1}(t)$, $v_{ds4}(t)$, $v_{ds5}(t)$, and $v_{ds8}(t)$, are clamped to 0. S_1, S_4, S_5, and S_8 are turned on at t_1, achieving ZVS of S_1, S_4, S_5, and S_8.

Power is transferred from the primary side to the load as L_{r1}, C_{r1}, L_{r2}, and C_{r2} start to resonate. S_1, S_4, S_5, and S_8 will turn off when the turn-on time reaches $T_r/2$, where T_r is the resonance period between L_{r1} and C_{r1}. Mode 1 ends.

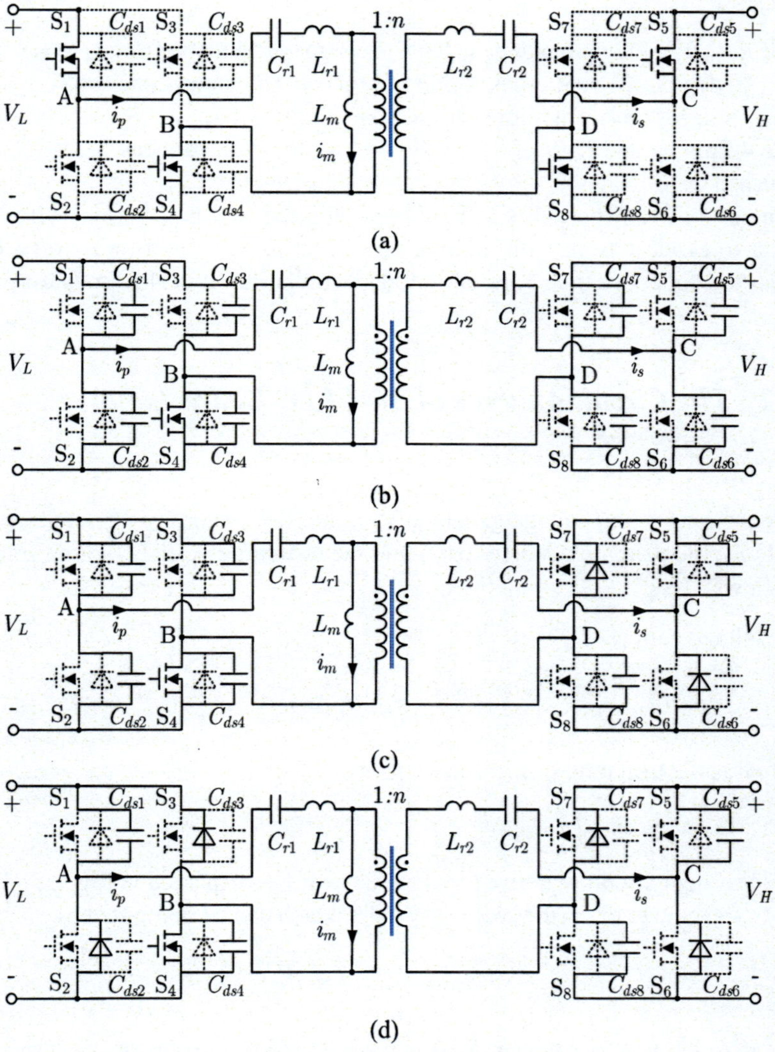

Fig. 6.4 Equivalent circuit: **a** Mode 1; **b** Mode 2; **c** Mode 3; **d** Mode 4

Mode 2 (t_2, t_3): Figure 6.3 depicts the Mode 2 waveforms (from t_2 to t_3), and Fig. 6.4b depicts the Mode 2 equivalent circuit. Between L_{r1}, C_{r1}, L_{r2}, C_{r2}, L_m, and all the parasitic capacitances of the switches, a complicated resonance starts to develop. While the voltages of S_5 and S_8 will rise, those of S_6 and S_7 will fall. Mode 2 will stop when the body diodes of S_6 and S_7 turn on and the voltages of S_5 and S_8 reach V_H.

Mode 3 (t_3, t_4): The equivalent circuit of Mode 3 is depicted in Fig. 6.4c and the waveforms of Mode 3 are presented in Fig. 6.3 (from t_3 to t_4). The body diodes of S_6 and S_7 are on in Mode 3. S_1 and S_4 will have an increase in voltage, whereas S_2 and S_3 will experience a drop. The body diodes of S_2 and S_3 will turn on when their voltages are 0. Mode 3 ends.

Mode 4 (t_4, t_5): Figure 6.3 depicts the waveforms of Mode 4 (from t_4 to t_5), whereas Fig. 6.4d depicts the analogous circuit of Mode 4. The body diodes of S_2, S_3, $S_{6,}$ and S_7 are on in Mode 4. Therefore, the voltages of S_2, S_3, $S_{6,}$ and S_7 are clamped to 0. Mode 4 finishes and ZVS of S_2, S_3, $S_{6,}$ and S_7 are attained when S_2, S_3, $S_{6,}$ and S_7 are turned on.

The second half period is symmetrical with the first half period, with Mode 5 similar to Mode 1 from t_5 to t_6; Mode 6 similar to Mode 2 from t_6 to t_7, Mode 7 similar to Mode 3 from t_7 to $t_{8,}$ and Mode 8 similar to Mode 4 from t_8 to t_9.

6.2.2 The Circuit Analysis of the CLLC Bidirectional Converter

The equations of the voltage gain (V_H/V_L), the RMS values of $i_p(t)$ and $i_s(t)$, $I_{p,rms}$ and $I_{s,rms}$; the peak flux density (B_{pk}) and the switching period, T_s are summarized as follows:

(a) Voltage gain [18]

$$V_H/V_L = n \tag{6.1}$$

where n is the turn ratio of the transformer.

(b) RMS values of $i_p(t)$: $I_{p,rms}$

The equivalent circuit to derive the impedance Z_{in} is displayed in Fig. 6.5. R_H is the load of the converter on the secondary side, which can be computed by

$$R_H = 8V_H^2/\left(\pi^2 P_H\right) \tag{6.2}$$

where P_H is the output power. R_{eqH} is the load referred to the primary side, which can be computed by

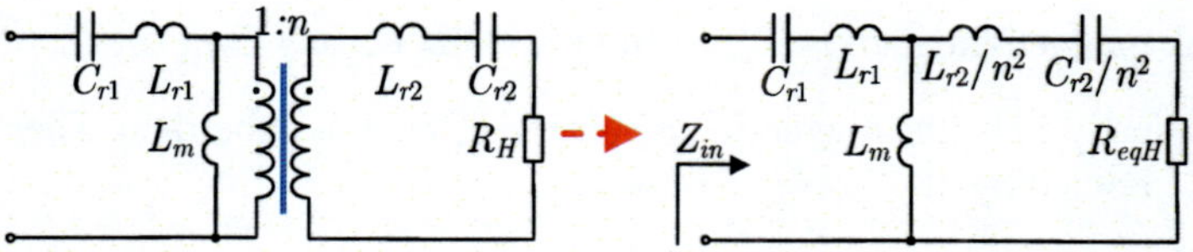

Fig. 6.5 Derivation of equivalent input impedance Z_{in}

$$R_{eqH} = 8V_H^2/\left(\pi^2 n^2 P_H\right) \tag{6.3}$$

In a symmetrical *CLLC* resonant converter,

$$L_{r2} = L_{r1}/n^2 \tag{6.4a}$$

$$C_{r2} = n^2 C_{r1} \tag{6.4b}$$

The impedance, Z_{in}, as shown in Fig. 6.5, can be computed by

$$Z_{in} = \sqrt{R_{eq}^2 + X_{eq}^2} \tag{6.5}$$

where R_{eq} is the real part and X_{eq} is the imaginary part. R_{eq} and X_{eq} can be computed by

$$R_{eq} = \omega_s^4 R_{eqH} L_m^2 C_{r1}^2 / \omega_s^2 R_{eqH}^2 C_{r1}^2 + \left(\omega_s^2 C_{r1} L_{r1} + \omega_s^2 C_{r1} L_m - 1\right)^2 \tag{6.6a}$$

$$X_{eq} = \frac{C_{r1}^3 L_{r1}(L_{r1} + L_m)(L_{r1} + 2L_m)\omega_s^6 + \sigma_2 \omega_s^4 + \sigma_3 \omega_s^2 - 1}{\omega_s C_{r1}\left[\omega_s^2 R_{eqH}^2 C_{r1}^2 + \left(\omega_s^2 C_{r1} L_{r1} + \omega_s^2 C_{r1} L_m - 1\right)^2\right]} \tag{6.6b}$$

where ω_s is the angular frequency, corresponding to f_s. σ_2 and σ_3 can be computed by

$$\sigma_2 = C_{r1}^2\left[(L_{r1} + L_m)\left(R_H^2 C_{r1} - L_{r1} - L_m\right) - 4L_{r1}L_m - 2L_{r1}^2 - L_m^2\right] \tag{6.7a}$$

$$\sigma_3 = 3C_{r1}(L_{r1} + L_m) - R_H^2 C_{r1}^2 \tag{6.7b}$$

$I_{p,rms}$ can be computed by

$$I_{p,rms} = V_L/\sqrt{2}Z_{in} \tag{6.8}$$

$I_{s,rms}$ can be computed by

$$I_{s,rms} = V_L/\sqrt{2}nZ_{in} \tag{6.9}$$

(c) Peak magnetizing current $I_{m,pk}$, and peak flux density B_{pk}

Under a steady state, the magnetizing current, $i_m(t)$, is symmetrical. Therefore, $I_{m,pk}$ can be computed by

$$I_{m,pk} = (V_L T_s)/(4L_m) \tag{6.10}$$

The peak magnetic flux density, B_{pk}, which will be utilized to compute the core loss of the transformer can be computed by

$$B_{pk} = \left(L_m I_{m,pk}\right)/\left(n_p A_e\right) \tag{6.11}$$

where n_p is the number of turns in the primary winding, A_e is the cross-sectional area of the magnetic core.

(d) Switching period (T_s)

Based on the analysis of the operating principles, the switching period of the *CLLC* resonant converter is

$$T_s = T_r + 2T_d \tag{6.12}$$

where T_d is the dead time ($T_d = t_5 - t_2$).

6.2.3 Conditions of ZVS and ZCS

As displayed in Fig. 6.3, before S_1, S_4, S_5 and S_8 are turned on, $i_p(t)$ and $i_s(t)$ are discharging C_{ds1}, C_{ds4}, C_{ds5}, and C_{ds8}, which makes the ZVS condition of S_1, S_4, S_5, and S_8. Similarly, S_2, S_3, S_6, and S_7 can also work under ZVS conditions because of $i_p(t)$ and $i_s(t)$. The dead time considering ZVS condition is [10]:

$$T_d \geq 16C_{oss} f_s L_m \tag{6.13}$$

where C_{oss} is the output capacitance of the switches. T_d is chosen as the lower limiting value of the aforementioned inequation in order to accomplish ZVS while maintaining high efficiency, which is:

$$T_d = 16C_{oss} f_r L_m \tag{6.14}$$

In order to achieve the soft commutation of the rectifiers, the duration of Mode 3 and 4 should be reserved, as a result,

$$f_s \leq f_r \tag{6.15}$$

It is important to note that the switching loss is decreased with the ZVS of the primary switches and the rectifiers. The driving loss, switch conduction loss, the power loss from resonant capacitances, copper loss, and transformer core loss are all still included in the overall power loss. As a result, further optimization is required for the converter's total power loss. As a result, in Sect. 6.3, a two-stage optimized technique will be used to reduce the total power loss of the *CLLC* resonant converter.

6.3 The Proposed AI Based High Efficiency Oriented Two-Stage Optimal Design Method for CLLC Bidirectional Power Converters in the Hybrid AC/DC Microgrid

The total power loss optimization of the open-loop *CLLC* resonant converter is presented in this part using a two-stage optimal design method based on an AI algorithm. In the first section, the equation for the total power loss in the *CLLC* resonant converter is derived. The ideal transformer parameters—the first step in the optimization process—are derived in the second portion using the presented AI based two-stage optimal design method. The planar transformer is designed in the third section using a hybrid electromagnetic analysis, with the leakage inductances serving as the resonant inductances obtained from the AI algorithm, which is the second stage of the optimization process.

6.3.1 Preliminary of the Proposed AI Based Two-Stage Optimal Design Method: Total Power Loss Equation of the CLLC Bidirectional Converter

Although ZVS of the switches is achieved, the total power loss (P_{tot}) in the *CLLC* resonant converter is still comprised of the driving loss of the main switches (P_{M_dr}) and the rectifiers (P_{R_dr}), the conduction loss of the main switches (P_{M_on}) and the rectifiers (P_{R_on}), copper loss of the transformer (P_{T_Cu}), core loss of the transformer (P_{T_Fe}) and the power loss of the resonant capacitance (P_{Cr}). Consequently, the P_{tot} can be written as

$$P_{tot} = P_{M\_dr} + P_{R\_dr} + P_{M\_on} + P_{R\_on} + P_{T\_Cu} + P_{T\_Fe} + P_{Cr} \tag{6.16}$$

The detailed expression of P_{M_dr}, P_{R_dr}, P_{M_on}, P_{R_on}, P_{T_Fe}, P_{T_Cu} and P_{Cr} can be derived as follows:

(a) Driving loss of main switches (rectifiers)

The driving loss can be computed by

$$P_{M(R)\_dr} = 4Q_{M(R)g} V_{M(R)gs} f_s \tag{6.17}$$

where $Q_{M(R)g}$ and $V_{M(R)gs}$ are the gate charge and the driving voltage of the main switches or rectifiers.

(b) **Conduction loss of main switches (rectifiers)**

The conduction loss can be computed by

$$P_{M(R)\_on} = 2I_{p(s),rms}^2 R_{M(R)\_on} \tag{6.18}$$

where $R_{M(R)\_on}$ is the on-resistance of the main switches (rectifiers). $I_{p(s),rms}$ is the corresponding RMS value of the resonant current, which can be found in (6.8) and (6.9).

(c) Copper loss of the transformer

The copper loss of the transformer can be computed by:

$$P_{T\_Cu} = I_{p,rms}^2 R_{T\_ac} \tag{6.19}$$

where R_{T_ac} is the AC resistance of the transformer referred to as the primary winding.

(d) The core loss of the transformer (P_{T_Fe})

The core loss can be computed based on the Steinmetz equation

$$P_{T\_Fe} = k_c f_s^{\alpha} (B_{pk})^{\beta} V_e \tag{6.20}$$

where B_{pk} can be found by (6.11). The datasheet of the magnetic material contains the values of k_c, α, and β. V_e is the volume of the core.

(e) The power loss of the resonant capacitances

The equivalent serial resistor (ESR) of the resonant capacitance can be computed by

$$R_{Cr1(2)} = \tan\delta / \left(2\pi f_s C_{r1(2)}\right) \tag{6.21}$$

where δ is the dielectric loss angle. Therefore, the power loss of the resonant capacitance can be computed by

$$P_{Cr1(2)} = I_{p(s),rms}^2 \tan\delta / \left(2\pi f_s C_{r1(2)}\right) \tag{6.22}$$

According to the analysis above, combing (6.17), (6.18), (6.19), (6.20), and (6.22), the total power loss of a *CLLC* resonant converter is

$$P_{tot} = 4Q_{Mg} V_{Mgs} f_s + 4Q_{Rg} V_{Rgs} f_s + I_{p,rms}^2 \left[2R_{M\_on} + R_{T\_ac} + \tan\delta/(2\pi f_s C_{r1})\right]$$

$$+ I_{s,rms}^2\left[2R_{R\_on} + \tan\delta/(2\pi f_s C_{r2})\right] + k_c f_s^{\alpha}\left[\left(L_m I_{m,pk}\right)/\left(n_p A_e\right)\right]^{\beta} V_e \tag{6.23}$$

6.3.2 The Proposed AI Based (GA + PSO) Two-Stage Optimal Design Methodology for the CLLC Bidirectional Converter

The flowchart of the proposed two-stage optimal design method is depicted in Fig. 6.6. The magnetic core is chosen at the outset of the proposed two-stage optimal design based on the power rating and switching frequency. The two stages carried out by the presented optimal design method are as follows:

Stage 1: Total power loss optimization based on the presented AI algorithm (GA + PSO)

The presented GA + PSO algorithm is utilized to minimize the total power loss and obtain the ideal L_{r1}, C_{r1}, L_{r2}, C_{r2}, $L_{m,}$ and P_{tot}, which will be used to construct the transformer.

Stage 2: Design of the transformer to obtain the ideal L_{r1}, L_{r2}, and L_m derived from the AI algorithm based on hybrid electromagnetic analysis

When the AI algorithm (GA + PSO) is used to derive L_{r1}, C_{r1}, L_{r2}, C_{r2}, $L_{m,}$ and $P_{tot,}$ appropriate d_w and d_a are applied in the transformer to obtain the ideal L_{r1}, L_{r2} and L_m.

Stage 1 and Stage 2 will be elaborated as follows:

(a) First Optimization Stage: Utilizing AI algorithm (GA + PSO) to optimize P_{tot}.

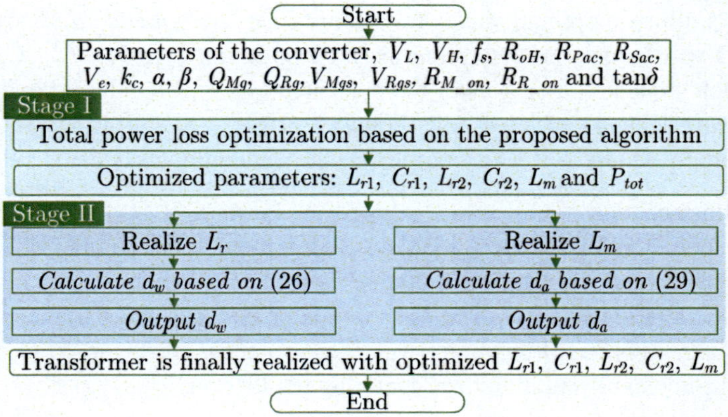

Fig. 6.6 Proposed two-stage optimal design methodology

(a1) Operation principles of the proposed AI algorithm (GA + PSO) in the optimal design.

The total power loss of the *CLLC* resonant converter is optimized in this section using the updated AI algorithm. The global optimum and the convergence iteration number conflict in both GA and PSO [21]: a better global optimum is always associated with a higher convergence iteration number, whereas a worse global optimum is always associated with a lower convergence iteration number. Furthermore, the number of iterations and the goal function are not optimized because the initial parameters and issue representation are always chosen based on user experience. Consequently, to obtain the needed accuracy in the experience-based GA or PSO, the iteration number is compromised. Four parameters—*Weight.start*, *Weight.end*, *kind* and *Vel.max*—are crucial in determining the effectiveness of an algorithm like PSO. PSO will be lacking if these four parameters are chosen at random.

The problem of a high number of iterations and calculation time, which is brought on by the random selection of *Weight.start*, *Weight.end*, kind and *Vel.max*, is addressed in this chapter via the hybrid AI algorithm GA + PSO. Figure 6.7 depicts the flowchart of the presented GA + PSO algorithm. The summary of the presented GA + PSO algorithm is depicted in Fig. 6.7a. The specifics of the GA part and PSO part are shown in Fig. 6.7b, c, respectively.

The presented GA + PSO algorithm, which is seen in Fig. 6.7a, can be thought of as an enhanced PSO in which the GA optimizes the PSO parameters. The GA is the starting point of the GA + PSO algorithm. First of all, the genes of each individual in GA, which are also the parameters of PSO (*Weight.start*, *Weight.end, kind* and *Vel.max*), are sent to the PSO as the algorithm parameters. The PSO algorithm is then executed given *Weight.start*, *Weight.end*, *kind,* and *Vel.max* from GA to extract the number of iterations N and total power loss P_{tot}, which is then returned to the GA algorithm. The fitness value of each individual in GA is evaluated using N and P_{tot}, and the parameters of PSO, *Weight.start*, *Weight.end*, *kind* and *Vel.max* are optimized and passed back to PSO. If the maximum number of iterations is reached after numerous iterations of this kind between the GA and PSO, the GA determines the PSO's ultimate optimal values for *Weight.start*, *Weight.end*, *kind,* and *Vel.max*. The PSO will be used to determine the final optimum L_{r1}, C_{r1}, L_{r2}, C_{r2}, $L_{m,}$ and P_{tot}, which are also the outputs of the presented GA + PSO algorithm, using ideal *Weight.start*, *Weight.end*, *kind,* and *Vel.max*.

The proposed GA + PSO accomplishes the sequential tasks listed below:

Step 1: Initialization of GA

The parameters of GA are initialized, which include the number of individuals, the crossover rate, the starting mutation rate, the ending mutation rate, and the maximum iteration number. The GA optimizer encodes the parameters of PSO, *Weight.start*, *Weight.end*, *kind,* and *Vel.max*, as genes, creates a string of genes to form a chromosome and initializes a starting population.

Step 2: Invoke PSO to obtain the number of iterations and best P_{tot} (PSO)

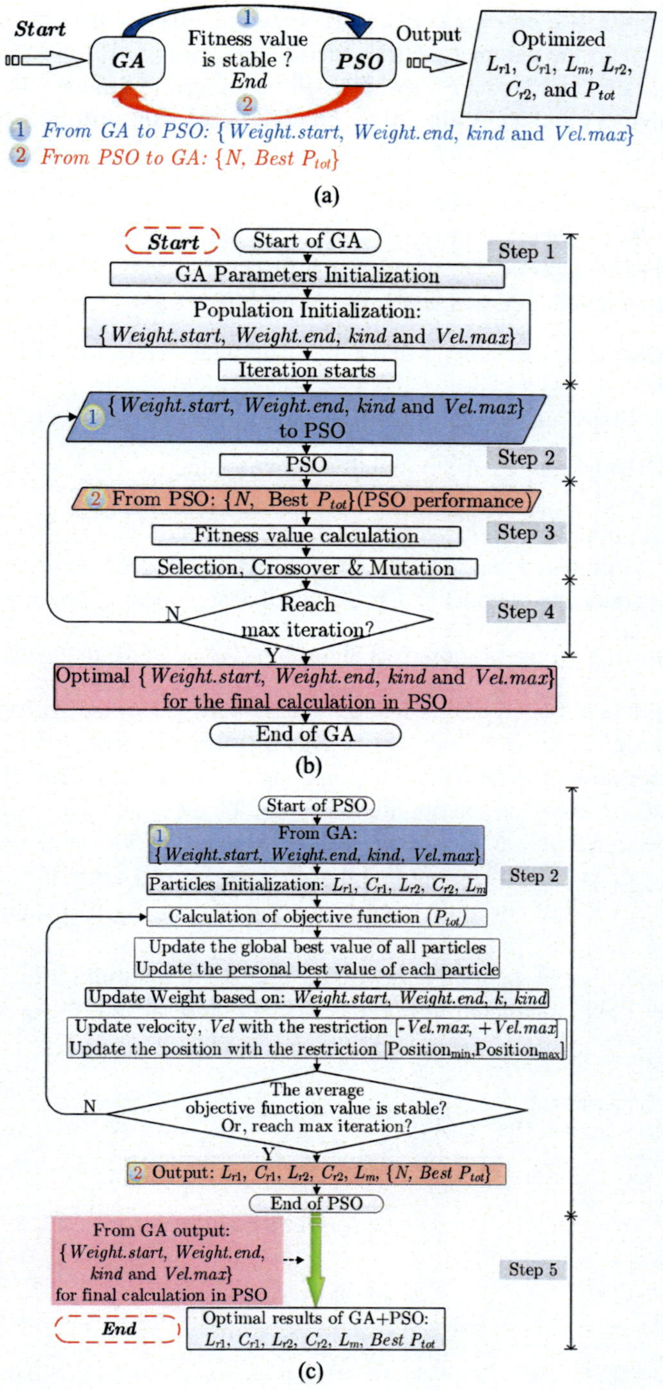

Fig. 6.7 Proposed recycling GA + PSO: **a** Structure diagram; **b** Flowchart of GA; **c** Flowchart of PSO

Weight.start, *Weight.end*, *kind,* and *Vel.max*, are transferred from GA to the PSO algorithm. With the parameters from each individual in GA, the PSO algorithm is performed to extract *N* and P_{tot}, which will be utilized to evaluate the fitness value of the individual in the GA algorithm. Figure 6.7c contains information about PSO in more detail.

Step 3: Evaluation and Evolution (GA)

N and P_{tot}, which are obtained from PSO, are utilized in GA to assess each individual's fitness value. The population in GA evolves through selection, crossover, and mutation based on the fitness value derived from PSO.

Step 4: Repeat Step 2 and Step 3 to obtain the optimal *Weight.start*, *Weight.end*, *kind,* and *Vel.max*

Step 2 to Step 3 are repeated until the maximum iteration number of GA is reached.

Step 5: Calculations of the optimal design parameters for the *CLLC* resonant converter (PSO)

With the optimum *Weight.start*, *Weight.end*, *kind,* and *Vel.max*, the *CLLC* resonant converter is optimized and ideal L_r, C_s, L_m, C_p, and P_{tot} are derived. When all the optimum parameters are obtained, the optimal design method is completed.

(a2) Evaluation of the proposed AI algorithm (GA + PSO) in the optimal design.

An example is provided in this section to assess the presented AI based two-stage optimal design method. Table 6.1 contains a list of the parameters for the *LCLC* resonant converter, the magnetic core, and the main switches. The proposed GA + PSO is used to compute the optimum L_{r1}, L_{r2}, C_{r1}, C_{r2} and P_{tot}. P_{tot} is appointed as the objective function. The range of L_{r1} is from 10 to 50 μH; the range of L_m is from 100 μH to 2 mH. The optimal design is performed using GA, PSO, and the presented GA + PSO in order to compare them to the single GA, and single PSO algorithms.

The fluctuation of the total loss with the number of iterations utilizing GA, PSO, and GA + PSO is depicted in Fig. 6.8. As can be observed, the presented GA +

Table 6.1 Parameters of *CLLC* resonant converter for validation

Parameters of the main circuit			
V_L	200 V	f_s	90 kHz
V_H	200 V	P_H	500 W
Magnetic core (ELP 64/10/50, N87)			
V_e	83,000 mm^3	α	4.823
A_e	1038 mm^2	β	5.521
k_c	3.716 × 10^{-24}		4.823
Main switches and rectifiers (C2M0080120D)			
V_{gs}	20.0 V	R_{on}	80 mΩ
Q_g	62 nC	C_{oss}	80 pF

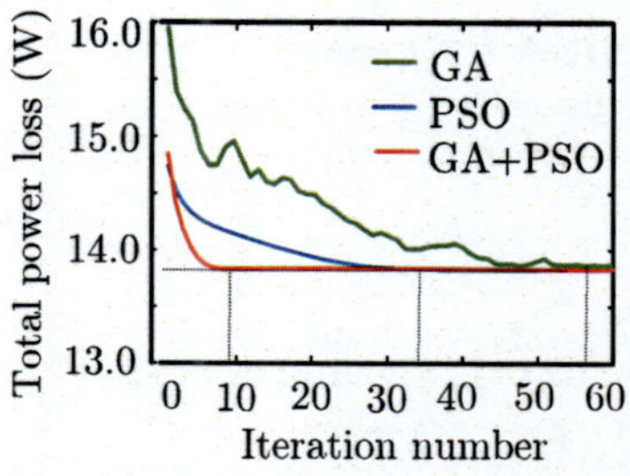

Fig. 6.8 Total power loss comparison

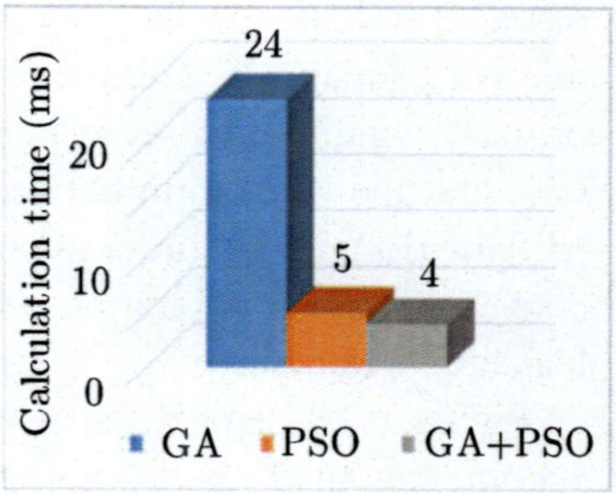

Fig. 6.9 Calculation time comparison

PSO and the other two optimal design methods (GA and PSO) both converge to the identical value of 13.76 W. Nevertheless, the GA, PSO, and proposed GA + PSO iteration numbers are 57, 33 and 9, respectively. In light of this, it can be said that the presented GA + PSO algorithm requires a lot fewer iterations than the other two algorithms.

In addition to the iteration number, the calculation time of GA, PSO, and proposed GA + PSO is compared in Fig. 6.9. It can be concluded that the proposed GA + PSO calculation time is shorter than the conventional GA and PSO.

The presented GA + PSO method performs better than the traditional GA or PSO algorithm, according to the comparison of iteration number and calculation time made above.

(a3) Final Optimal results of L_{r1}, L_{r2}, L_m, C_{r1}, C_{r2} and P_{tot}

The ideal L_{r1}, L_{r2}, L_m, C_{r1}, C_{r2}, and P_{tot} can be derived using the aforementioned AI (GA + PSO) method and are displayed in Table 6.2. Table 6.2 shows that $L_{r1} = 14.8\ \mu\text{H}$, $C_{r1} = 196.0\ \text{nF}$, $L_m = 1.0\ \text{mH}$, and that $P_{tot} = 13.76\ \text{W}$ is the ideal power loss. It is important to note that C_{r1} and C_{r2} can easily be built to have values of 196.0 nF because they are discrete capacitors. Nevertheless, the parasitic parameters of the transformer, L_{r1}, L_{r2}, and L_m, are governed by the winding arrangements of the transformer. As a result, a transformer will be developed in the section that follows the ideal parasitic parameters.

(b) Second optimization stage: optimal design of the transformer to obtain the optimal parasitic parameters derived from AI algorithm via hybrid electromagnetic analysis

Table 6.2 Optimal parameters of the *CLLC* resonant converter

L_{r1}	14.8 μH	C_{r1}	196.0 nF
L_{r2}	14.8 μH	C_{r2}	196.0 nF
L_m	1.0 mH		

(b1) Derivation of the leakage inductances

Power converters are evolving in the present day toward high efficiency and high power density. The leakage inductances of the planar transformer are used as the resonant inductances in this chapter to boost the power density while lowering the number of magnetic components (L_{r1} and L_{r2}). In this section, the thickness of the primary winding and secondary winding, d_p and d_s, the thickness of the insulation (d_i), and the separation between the primary winding and the secondary winding (d_w) are designed in order to obtain the optimal L_{r1} and L_{r2}. For the *CLLC* resonant converter, a generic equation is derived to determine d_w in order to attain the necessary leakage inductances.

Figure 6.10 depicts the winding configuration of a planar transformer and its accompanying magneto-motive force (MMF). The primary winding has n_p turns dispersed in m_p layers, whereas the secondary winding contains n_s turns spread in m_s layers. The distance between the main and secondary windings is d_w, and it will be determined to obtain the best leakage inductances. To get the ideal magnetizing inductances, the air gap (d_a) is added.

Assuming the current excitation to the primary winding is I_p, based on the distribution MMF in Fig. 6.10, the magnetic energy stored in the core window can be computed by

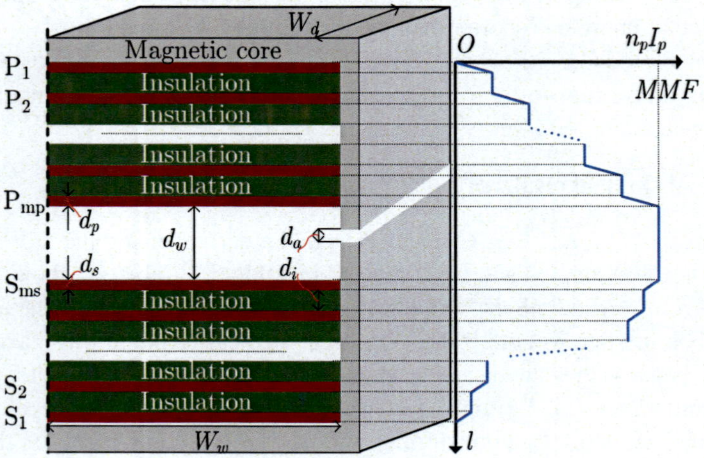

Fig. 6.10 General winding configuration and MMF distribution

$$E_m = \frac{\mu_0 \mu_r n_p^2 I_p^2 W_d}{6W_w}\left[2m_p d_p + 2m_s d_s + \frac{(m_p - 1)(2m_p - 1)}{m_p}d_i + \frac{(m_s - 1)(2m_s - 1)}{m_s}d_i + 6d_w\right] \tag{6.24}$$

where W_d is the depth of the window, W_w is the width of the window. By the relation between leakage inductance and magnetic energy [22], leakage inductances can be computed by (6.25)

$$L_{lk1} = L_{lk2} = \frac{\mu_0 \mu_r n_p^2 W_d}{6W_w}\left[2m_p d_p + 2m_s d_s + \frac{(m_p - 1)(2m_p - 1)}{m_p}d_i + \frac{(m_s - 1)(2m_s - 1)}{m_s}d_i + 6d_w\right] \tag{6.25}$$

Therefore, the optimum leakage inductances can be obtained by modifying d_w, according to the result in (6.25).

Here, d_w is the distance between the primary winding and the secondary winding, which is utilized to obtain the ideal leakage inductance and can be computed by (6.26)

$$d_w = \frac{W_w L_{lk1}}{\mu_0 \mu_r n_p^2 W_d} - \frac{m_p d_p}{3} - \frac{m_s d_s}{3} - \frac{(m_p - 1)(2m_p - 1)}{6m_p}d_i - \frac{(m_s - 1)(2m_s - 1)}{6m_s}d_i \tag{6.26}$$

(b2) Derivation of the magnetizing inductance

The thickness of the air gap (d_a) is changed to obtain the desired magnetizing inductance. This section develops the equation needed to determine the air gap's thickness (d_a) in order to attain the appropriate magnetizing inductance (L_m). Figure 6.11 displays the corresponding equivalent magnetic circuit and 3D model of a planar transformer with an air gap.

The magnetic reluctances, R_{g1}, R_{g2}, R_{g3}, R_{c1}, and R_{c2} can be computed by

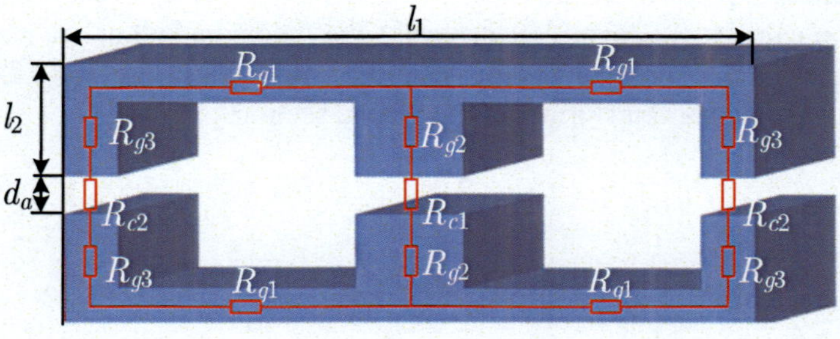

Fig. 6.11 Equivalent magnetic circuit of a planar transformer with an air gap

$$R_{g1} = l_1/(\mu_0\mu_r A_e) \tag{6.27a}$$

$$R_{g2} = l_2/(\mu_0\mu_r A_e) \tag{6.27b}$$

$$R_{g3} = 2l_2/(\mu_0\mu_r A_e) \tag{6.27c}$$

$$R_{c1} = d_a/(\mu_0 A_e) \tag{6.27d}$$

$$R_{c2} = 2d_a/(\mu_0 A_e) \tag{6.27e}$$

where the size of the magnetic core, shown in Fig. 6.11, is l_1 and l_2. A_e is the cross-sectional area of the center leg.

Therefore, the magnetizing inductance can be computed by

$$L_m = \mu_0\mu_r A_e n_p^2/(l_1 + 4l_2 + 3\mu_r d_a) \tag{6.28}$$

From (6.28), it is clear that the magnetizing inductances will decrease as d_a increases. Consequently, the ideal L_m can be obtained by altering d_a.

The thickness of the air gap, d_a, which is utilized to obtain the ideal L_m, can be computed by

$$d_a = \left(\mu_0 A_e n_p^2\right)/(3L_m) - (l_1 + 4l_2)/(3\mu_r) \tag{6.29}$$

6.4 Simulation and Experimental Validations of the Proposed AI Based Two-Stage Optimal Design Method

In this section, simulations and experiments are used to support the presented AI-based optimal design method. Building the planar transformer for the *CLLC* resonant converter takes place in the first section using the d_w and d_a equations. In order to verify the calculations, hybrid electromagnetic field simulations are also run. The proposed two-stage optimal design method is evaluated using the *CLLC* resonant converter and planar transformer in the second section.

6.4.1 *Design of a Planar Transformer with the Desired Parasitic Parameters for a CLLC Bidirectional Converter*

The magnetic core, ELP 64/10/50, with N87 as the magnetic material from TDK is chosen, taking into account output power. The power capacity is increased by combining two ELP 64/10/50 to boost A_e. Table 6.3 is a list of the magnetic core's characteristics. The planar transformer for the *CLLC* resonant converter will be designed using the parameters in Table 6.3.

(a) Design of a planar transformer based on (6.26) and (6.29)

In order to obtain the optimized parameters listed in Table 6.2, d_w and d_a can be computed by (6.26) and (6.29), which are

$$d_w = 9.15\,\text{mm} \tag{6.30a}$$

$$d_a = 0.14\,\text{mm} \tag{6.30b}$$

The hybrid electromagnetic simulations will validate the calculated d_w and d_a.

(b) Hybrid electromagnetic field simulation validation

In Ansys Maxwell, the planar transformer model with estimated d_w and d_a is constructed, and hybrid electromagnetic field simulations are run to verify the accuracy of the calculations. Figure 6.12a, b, respectively, display the distributions of magnetic flux density used to compute the leakage inductances and magnetizing inductance. The simulated L_{r1}, $L_{r2,}$ and L_m are displayed in Table 6.4 based on Fig. 6.12.

As displayed in Fig. 6.13, the planar transformer is constructed using the calculated d_w and d_a for the *CLLC* resonant converter.

In addition, the impedance analyzer also measures L_{r1}, $L_{r2,}$ and L_m, and the measured findings are displayed in Table 6.4. Table 6.4 shows that the calculated results and simulated outcomes are in agreement, validating the efficiency of (6.26) and (6.29). To further confirm the presented two-stage optimal design method, the planar transformer will be tested in the *CLLC* resonant converter in the next section.

Table 6.3 Parameters of the magnetic core

n_p	16	d_i	0.1 mm
n_s	16	W_w	22 mm
m_p	4	W_d	100 mm
m_s	4	A_e	1038 mm^2
d_p, d_s	70 μm	l_1	64 mm
μ_r	1490	l_2	10 mm

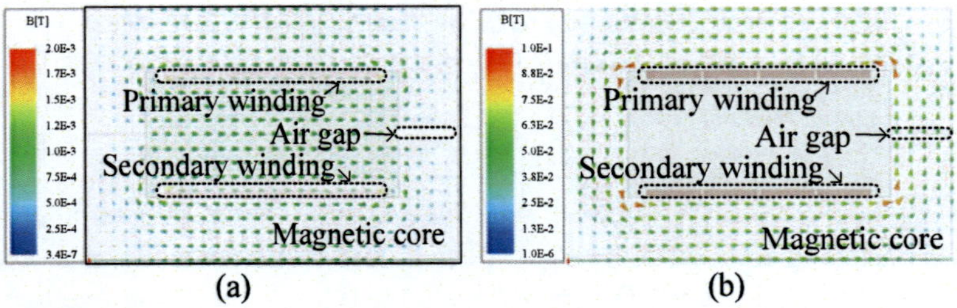

Fig. 6.12 **Distribution of magnetic flux density for calculations of: a** Leakage inductance; **b** Magnetizing inductance

Table 6.4 Comparison of leakage inductances and magnetizing inductance

Parameters	Optimal	Simulated	Measured
L_{r1} (μH)	14.8	14.38	14.73
L_{r2} (μH)	14.8	14.38	14.73
L_m (mH)	1.0	1.02	1.03

Fig. 6.13 The planar transformer for the *CLLC* resonant converter

6.4.2 *Experimental Validation of the Proposed AI Based Two-Stage Optimal Design Methodology*

The *CLLC* resonant converter is constructed to verify the presented AI-based two-stage optimum design method, as illustrated in Fig. 6.14, where the designed planar transformer is employed.

(a) Soft-switching verifications of the *CLLC* resonant converter

Experiments are conducted to demonstrate the ZVS of the switches, and Fig. 6.15 displays the driving signals of S_1 and S_5, $v_{gs1}(t)$ and $v_{gs5}(t)$, the primary current and the secondary current, $i_p(t)$ and $i_s(t)$, the voltages of S_1 and S_5, $v_{ds1}(t)$ and $v_{ds5}(t)$.

Figure 6.15a demonstrates that before S_1 is turned on, the voltage of S_1, $v_{ds1}(t)$, has fallen to 0. Consequently, when S_1 is turned on, S_1 achieves ZVS. Similar to S_1, Fig. 6.15b demonstrates that before S_5 is turned on, its voltage, $v_{ds5}(t)$, has dropped to 0. Therefore, S_5 achieves ZVS when it is turned on. The theoretical analysis and experimental results are in agreement.

(b) Efficiency verification of the AI based optimal design method

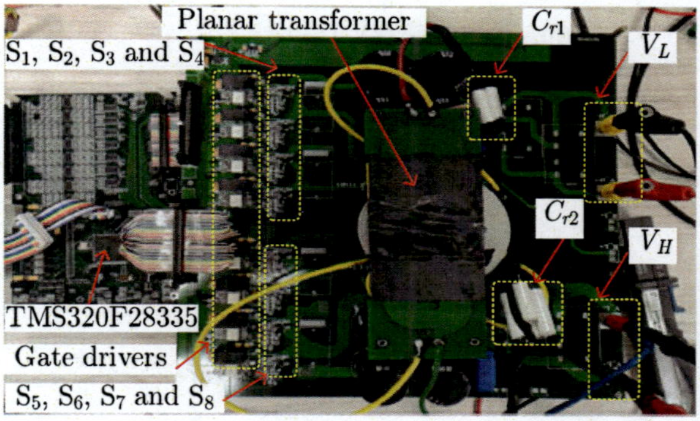

Fig. 6.14 The *CLLC* resonant converter for the validation

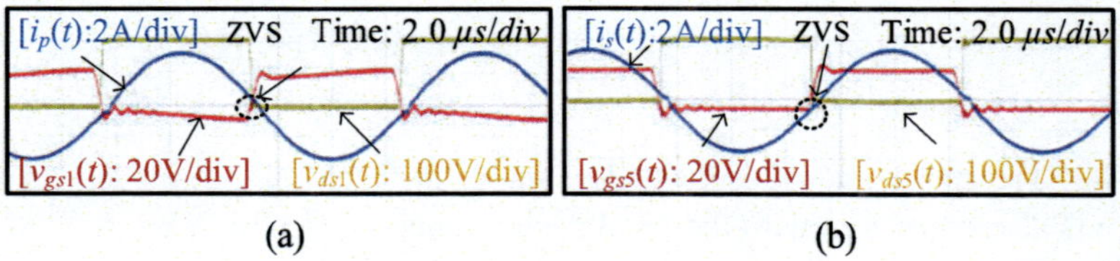

Fig. 6.15 Experimental waveforms of S_1 and S_5 to validate the soft-switching characteristics: **a** S_1; **b** S_5

In order to assess the efficiency of the presented optimal design method, L_{r1}, L_m and C_{r1} are adjusted and the corresponding efficiency is measured under rated output power. In Fig. 6.16, the measured effectiveness is depicted.

As shown in Fig. 6.16, L_{r1}, L_m, and C_{r1} all have an effect on the effectiveness of the *CLLC* resonant converter. The efficiency of the two-stage optimal design method proposed in this chapter is validated by the fact that effectiveness is higher under the ideal condition than in other circumstances.

(c) Efficiency under different output power

Figure 6.16 displays the waveforms of the input voltage and input current, V_L, i_L, and the output voltage and output current, V_H, i_H. As illustrated in Fig. 6.17, the efficiency of the *CLLC* resonant converter is assessed with various input powers. As observed in Fig. 6.18, the efficiency is up to 97.02% at the rated output power

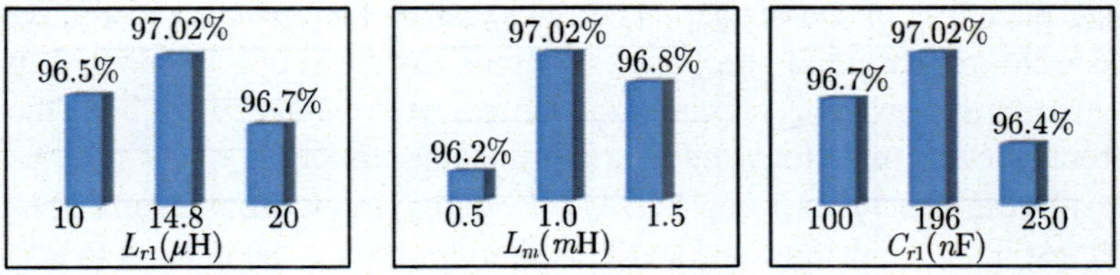

Fig. 6.16 Comparison of efficiency under different L_{r1}, L_m and C_{r1}

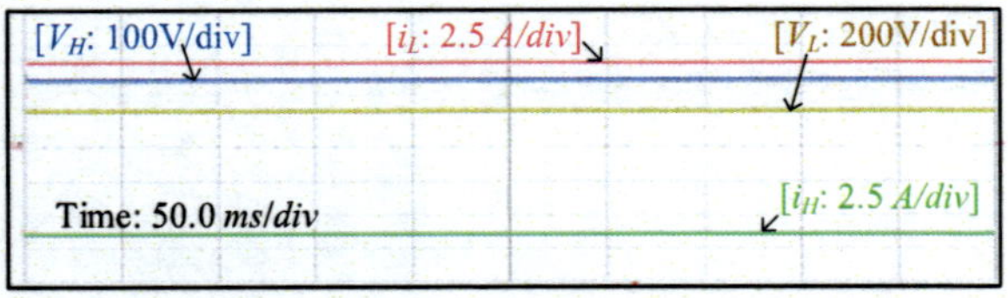

Fig. 6.17 Waveforms of V_L, V_H, i_L and i_H

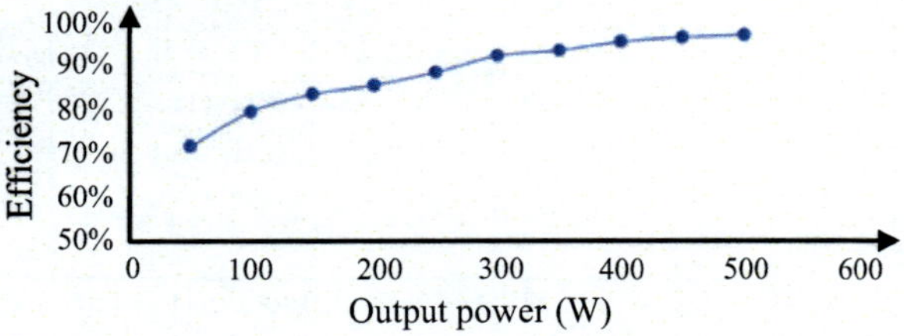

Fig. 6.18 The efficiency of the *CLLC* resonant converter versus output power

(500W). The measured total power loss under the rated output power is 14.92 W, which is also consistent with the ideal design outcome (13.76 W).

6.5 Conclusions

For applications involving hybrid AC/DC microgrids, a two-stage optimal design methodology for high-efficiency *CLLC* resonant converters based on AI (GA + PSO) has been presented. The proposed approach focuses on optimizing the magnetic design and overall power of the *CLLC* resonant converter using an AI algorithm. The driving loss, conduction loss of the switches, power loss of the resonant capacitances, copper loss of the transformer, and core loss of the transformer are all included in the total power loss that is optimized by the GA + PSO algorithm in the first stage. From this, the optimized L_{r1}, L_{r2}, C_{r1}, $C_{r2,}$ and L_m are derived. The planar transformer with ideal L_{r1}, $L_{r2,}$ and L_m is constructed in the second stage in order to accomplish the AI optimal design. An example real converter has been used to validate the proposed optimum design method.

The planar transformer design for the *CLLC* resonant converter has also been researched in addition to the optimal design method. It is proposed to use general equations to determine the distance between the primary and secondary windings and the thickness of the air gap. The thickness of the air gap and the separation between the primary and secondary windings can be simply computed using the presented equations. The presented equations were validated by both the hybrid electromagnetic simulations and the experimental results.

Despite the difficulties brought on by the various circuit variables and their internal couplings, it can be determined that the AI algorithm can be utilized to address the total power loss optimization. Additionally, It is also important to note that applying an AI algorithm to the optimum power converter design primarily involves two difficult steps: at the first, optimizing the power converter using AI algorithms to determine the ideal circuit parameters; secondly, implementing the ideal circuit parameters. For example, in this chapter, If the AI algorithm optimizes the leakage inductor to 14.8 μH, how to ensure that the real designed transformer leakage inductance matches the AI optimized value is just as crucial as the AI optimization itself.

References

1. Y. Xia, W. Wei, M. Yu, X. Wang and Y. Peng, "Power Management for a Hybrid AC/DC Microgrid with Multiple Subgrids,"*IEEE Trans. on Power Electronics*, vol. 33, no. 4, pp. 3520–3533, April 2018.
2. P. C. Loh, D. Li, Y. K. Chai and F. Blaabjerg, "Autonomous Operation of Hybrid Microgrid With AC and DC Subgrids," *IEEE Trans. on Power Electronics*, vol. 28, no. 5, pp. 2214–2223, May 2013.
3. Y. Xia, Y. Peng, P. Yang, M. Yu and W. Wei, "Distributed Coordination Control for Multiple Bidirectional Power Converters in a Hybrid AC/DC Microgrid," *IEEE Trans. on Power Electro.*, vol. 32, no. 6, pp. 4949–4959, 2017.
4. F. Nejabatkhah and Y. W. Li, "Overview of Power Management Strategies of Hybrid AC/DC Microgrid," *IEEE Trans. on Power Electronics*, vol. 30, no. 12, pp. 7072–7089, Dec. 2015.
5. N. Eghtedarpour and E. Farjah, "Power Control and Management in a Hybrid AC/DC Microgrid," *IEEE Trans. on Smart Grid*, vol. 5, no. 3, pp. 1494–1505, May 2014.
6. P. He and A. Khaligh, "Comprehensive Analyses and Comparison of 1 kW Isolated DC–DC Converters for Bidirectional EV Charging Systems," *IEEE Trans. on Transport. Electr.*, vol. 3, no. 1, pp. 147–156, 2017.
7. S. Zhao, Q. Li, F. C. Lee and B. Li, "High-Frequency Transformer Design for Modular Power Conversion From Medium-Voltage AC to 400 VDC," *IEEE Trans. on Power Electronics*, vol. 33, no. 9, pp. 7545–7557, 2018.
8. B. Zhao, Q. Song, W. Liu and Y. Sun, "Overview of Dual-Active-Bridge Isolated Bidirectional DC–DC Converter for High-Frequency-Link Power-Conversion System," *IEEE Trans. on Power Electronics*, vol. 29, no. 8, pp. 4091–4106, Aug. 2014.
9. J. H. Jung, H. S. Kim, M. H. Ryu and J. W. Baek, "Design Methodology of Bidirectional *CLLC* Resonant Converter for High-Frequency Isolation of DC Distribution Systems," *IEEE Trans. on Power Electronics*, vol. 28, no. 4, pp. 1741–1755, April 2013.
10. Z. U. Zahid, Z. M. Dalala, R. Chen, B. Chen and J. S. Lai, "Design of Bidirectional DC–DC Resonant Converter for Vehicle-to-Grid (V2G) Applications," *IEEE Trans. on Transpor. Electrifi.*, vol. 1, no. 3, pp. 232–244, 2015.
11. J. Huang, X. Zhang, Z. Shuai, X. Zhang, P. Wang, L. H. Koh, J. Xiao and X. Tong, "Robust Circuit Parameters Design for the *CLLC*-Type DC Transformer in the Hybrid AC/DC Microgrid," *IEEE Trans. on Industrial Electronics.* (early access).
12. C. Zhang, P. Li, Z. Kan, X. Chai and X. Guo, "Integrated Half-Bridge *CLLC* Bidirectional Converter for Energy Storage Systems," *IEEE Trans. on Industrial Electronics*, vol. 65, no. 5, pp. 3879–3889, May 2018.
13. W. L. Malan, D. M. Vilathgamuwa and G. R. Walker, "Modeling and Control of a Resonant Dual Active Bridge with a Tuned *CLLC* Network," *IEEE Trans. on Power Electronics*, vol. 31, no. 10, pp. 7297–7310, 2016.
14. S. Zou, J. Lu, A. Mallik and A. Khaligh, "Bi-Directional *CLLC* Converter With Synchronous Rectification for Plug-In Electric Vehicles," *IEEE Trans. on Industry Applications*, vol. 54, no. 2, pp. 998–1005, March-April 2018.

15. G. Liu, D. Li, J. Q. Zhang, B. Hu and M. L. Jia, "Bidirectional *CLLC* resonant DC-DC converter with integrated magnetic for OBCM application," *2015 IEEE International Conference on Industrial Technology (ICIT)*, Seville, 2015, pp. 946–951.
16. B. Li, Q. Li and F. C. Lee, "A WBG based three phase 12.5 kW 500 kHz *CLLC* resonant converter with integrated PCB winding transformer," 2018 IEEE Applied Power Electronics Conference and Exposition (APEC), San Antonio, TX, 2018, pp. 469–475.
17. S. Zhao, Q. Li, F. Lee and B. Li, "High Frequency Transformer Design for Modular Power Conversion from Medium Voltage AC to 400V DC," *IEEE Trans. on Power Electronics.* (early access).
18. X. Lai, P. Zhang, Y. Wang and M. Wu, "Position-Posture Control of a Planar Four-Link Underactuated Manipulator Based on Genetic Algorithm," *IEEE Trans. on Indus. Electr*, vol. 64, no. 6, pp. 4781–4791, 2017.
19. S. E. De León-Aldaco, H. Calleja and J. Aguayo Alquicira, "Metaheuristic Optimization Methods Applied to Power Converters: A Review," *IEEE Trans. on Power Electronics*, vol. 30, no. 12, pp. 6791–6803, Dec. 2015.
20. J. Huang, J. Xiao, C. Wen, P. Wang and A. Zhang, "Implementation of Bidirectional Resonant DC Transformer in Hybrid AC/DC Micro-grid," *IEEE Trans. on Smart Grid.* (early access).
21. H. Shi, H. Wen, Y. Hu and L. Jiang, "Reactive power Minimization in Bidirectional DC-DC converters Using A Unified-Phasor-Based Particles Swarm Optimization," *IEEE Trans. on Power Electro.*, vol. PP, no. 99, pp. 1–1.
22. B. Zhao, Z. Ouyang, M. Duffy, M. A. E. Andersen and W. G. Hurley, " An Improved Partially Interleaved Transformer Structure for High-voltage High-frequency Multiple-output Applications," *IEEE Trans. on Industrial Electronics.* (early access).

Chapter 7
The Proposed Artificial-Intelligence-Based Triple Phase Shift Modulation for Dual Active Bridge Converter with Minimized Current Stress

7.1 Introduction

Since it was first introduced in 1992 [1], the dual active bridge (DAB) isolated bidirectional DC-DC converter has been widely used in a variety of applications, including wireless power transfer, electric vehicles, DC microgrids [2], and solid state transformers [3]. One high-frequency transformer and two full bridges make up this topology, as depicted in Fig. 7.1. It draws a lot of interest due to its galvanic isolation, high power density, and ability to transfer power in both directions [4, 5].

Phase shift modulation is the most widely utilized modulation strategy for DAB converters due to its straightforward installation and fundamental frequency functioning [6]. Single phase shift (SPS) modulation, which has one degree of control freedom in the phase shift between two complete bridges, is the most basic phase shift modulation approach [7]. Extended phase shift (EPS), which modifies SPS modulation, increases the amount of controllable freedom in the duty cycle of a full bridge by one [8, 9]. Similar to SPS, dual phase shift (DPS) modulation shares the same value and makes the duty cycle of two full bridges programmable [10, 11]. Triple phase shift (TPS) modulation, which realizes three degrees of control flexibility overall [12–15], has been introduced to incorporate one additional degree of control freedom and further increase modulation performance. SPS, EPS, and DPS can be viewed as special examples of TPS because it has the ability to regulate the duty cycle of two full bridges as well as the phase shift between them. TPS modulation increases power efficiency and widens the zero voltage switching (ZVS) range when all modulation variables are used [16]. Consequently, TPS can be thought of as one of the most cutting-edge methods of phase shift modulation.

A common trend in the TPS modulation for DAB converter at the moment is the optimization of current stress. The peak inductor current is considered to be the current stress objective to be optimized in this chapter, which has benefits in many areas. First of all, peak current serves as a cap on switching losses, therefore maximizing it helps lower switching losses [17]. Peak current optimization can also decrease root-mean-square current and increase efficiency [18]. Additionally, the

X. Zhang et al., *Automated Design of Electrical Converters with Advanced AI Algorithms*,
https://doi.org/10.1007/978-981-99-0459-4_7

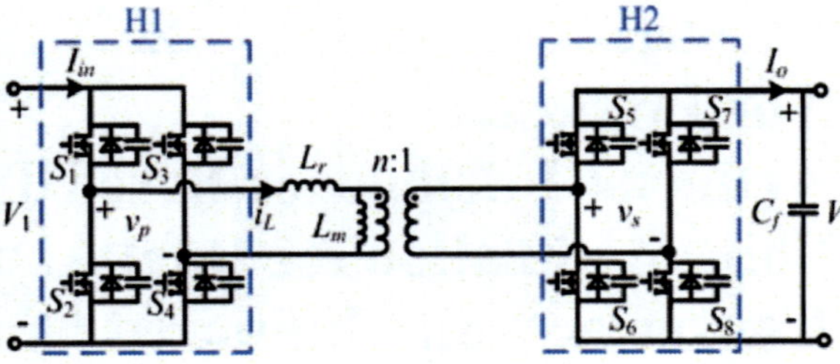

Fig. 7.1 Typical schematic of an isolated DAB converter with single inductor L

right amount of peak current stress can shrink the magnetic core and safeguard power electronics [19, 20].

However, the analytical formula for current stress is incredibly challenging to calculate [21, 22] because of the complexity imposed by the three degrees of control freedom in TPS modulation. In earlier studies, which took a lot of effort [6, 12, 22], researchers looked at the waveforms in each operating mode and piecemeal integrated the current. The complex analytical formula that is developed makes it difficult to optimize using these manually generated mathematical phrases. Before the derivation process, some assumptions are made in addition to the high computational complexity, such as the assumption of lossless components and a small magnetizing current [18, 23]. In addition, despite simplifying the analysis, these suppositions and approximations compromise the validity of the analytical results.

There are typically two approaches to implementing the TPS modulation strategy. One approach is to keep the modulator's deduced formula for optimal outcomes [18, 22, 23]. The outcomes of the modulation will be computed under realistic circumstances. As was already mentioned, while the method based on analytical formula is simple to use, it has a time-consuming problem and a difficult deduction procedure. The alternative strategy is to use a lookup table to store the outcomes of optimal modulation rather than generating analytical formulas [24–26]. The modulation variables in this lookup table will be searched for after the current operating conditions have been determined, after which they will be applied. The lookup table's modulation results, however, are discrete, which results in a situation where the practical requirements are not included in this table.

Researchers have thought about using various AI technologies to get over challenges in optimizing TPS modulation for DAB converters. In [26], Tang et al. made an effort to reduce human dependence during the optimization process. To locate the most effective modulation variables, which are kept in a lookup table, they used Q-learning. The lookup table produces answers for discrete modulation, but the analytical procedure to construct formulas with approximations continues. Additionally, in order to reduce reactive power, Harrye et al. utilized a neural network (NN) as a TPS controller [27]. However, because the NN controller is used in an open loop control, the reactive power still needs to be calculated manually, and the modulation performance is not as smooth as intended.

An AI-based TPS modulation (AI-TPSM) technique is proposed in this chapter with the aim of reducing the current stress on the DAB converter when it is subject to TPS modulation. AI-TPSM typically has three stages and uses three different AI tools. First, NN is trained using simulations to figure out the correlations between the variables under consideration and the current level of stress. This replaces the labor-intensive and erroneous old analytical process. Second, the particle swarm optimization (PSO) technique is used to identify the outcomes of the best modulation that can reduce the current stress. In order to provide continuous modulation, the optimal modulation results under various operating situations are stored using a fuzzy inference system (FIS). The proposed AI-TPSM has a high level of automation, which can lessen the workload for engineers and increase accuracy.

The organization of this chapter is as follows: after the introduction, operation principles of TPS modulation and current challenges are discussed in Sect. 7.2; the detailed process of AI-TPSM is illustrated in Sect. 7.3; an example of an application case for AI-TPSM is provided in Sect. 7.4; hardware experimental results are shown in Sect. 7.5, and the conclusion is summarized in Sect. 7.6.

7.2 Operating Principle of TPS Modulation and the Existing Challenges

7.2.1 Operating Principle of TPS Modulation for DAB Converter

Figure 7.1 shows the DAB converter's circuit design. A magnetic tank that incorporates an inductor L and a high-frequency transformer joins two complete bridges H_1 and H_2 together. H_1 and H_2 produce ac voltages of v_p and v_s, respectively. Figure 7.2 shows the gate driving signals and waveforms of v_p and v_s when TPS modulation is used.

$2T$ represents a whole switching period T_s. in Fig. 7.2. D_0 is the difference in phase between two full bridges that belong to [− 1, 1], S_1 and S_5. The duty cycles for v_p and v_s are D_1 and D_2, respectively. D_1 and D_2 both have [0, 1] as their ranges. The

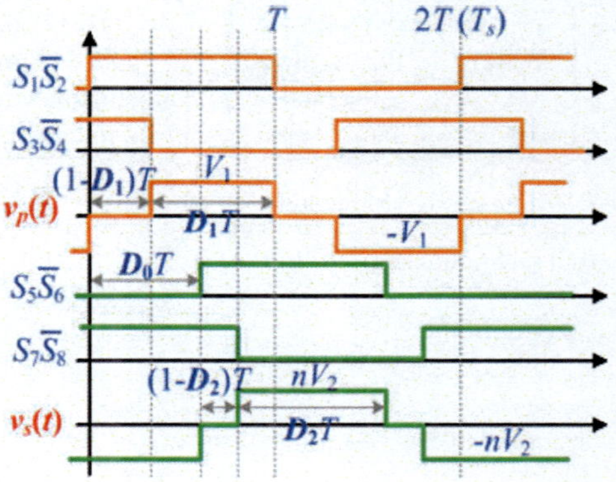

Fig. 7.2 Operating principles of TPS modulation

transmitted power and the inductor current $i_L(t)$ can both be controlled by adjusting any one of the three degrees of control freedom. The maximum power that can be transferred, in accordance with [22], can be stated as (7.1), where f_s is the switching frequency. The value of a single inductor L is determined by the maximum power to be transmitted, P_{max} (7.2).

$$P_{max} = \frac{nV_1V_2}{8f_sL} \tag{7.1}$$

$$L \leq \frac{nV_1V_2}{8f_sP_{max}} \tag{7.2}$$

7.2.2 *Challenge Descriptions for Optimization of TPS Modulation with Minimized Current Stress*

As shown in Fig. 7.3, there are normally three stages to the process of achieving optimal TPS modulation with reduced current stress and implementing this modulation in practice.

Following the definition of the operational parameters, the associations between the variables under stress and the current stress are examined. Variables include operating circumstances (output power P and output voltage V_2), modulation variables (D_0, D_1, and D_2), and current stress (expressed by the peak i_{pk} of the current through the inductor i_L). The modulation variables will then be adjusted for the least amount of current stress under various operating situations. The last stage will involve the real-time implementation of the improved modulation variables under various operational scenarios.

As illustrated in Fig. 7.4, there are several obstacles in Stage I and Stage III of this procedure, which heighten complexity and jeopardize accuracy.

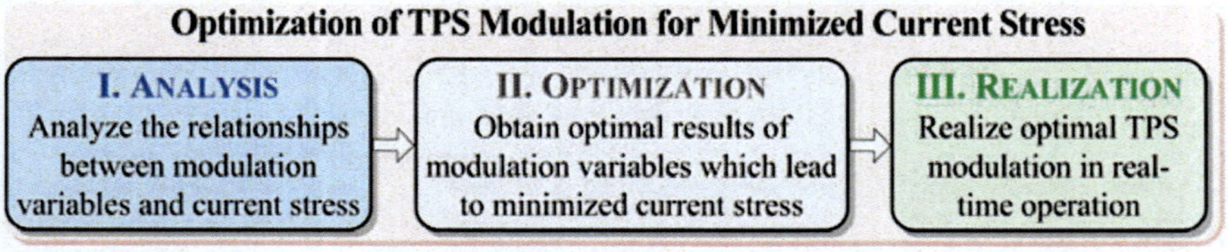

Fig. 7.3 The process to achieve optimal TPS modulation with minimized current stress and carry out the modulation

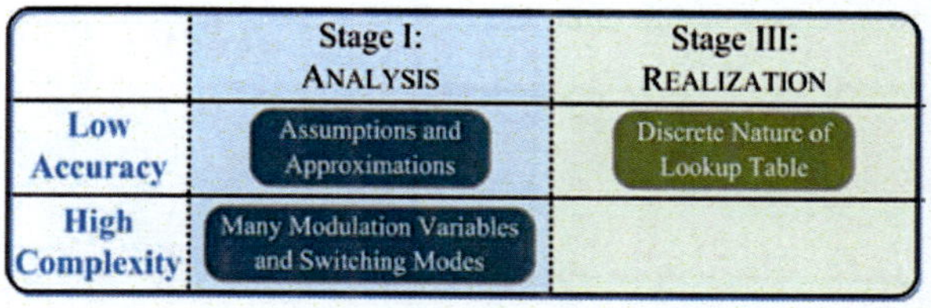

Fig. 7.4 Challenge descriptions for optimization of TPS modulation

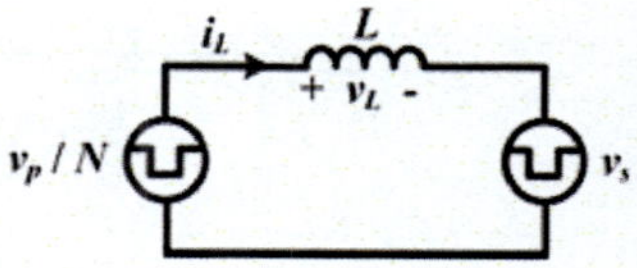

Fig. 7.5 Equivalent circuit of DAB converter with a single L

7.2.2.1 Challenge in Stage I: Analysis of Current Stress

The process described below is used to analyze current stress in earlier research studies [18, 19, 22, 23]. To investigate the inductor current $i_L(t)$ and its peak value i_{pk} (current stress) under all switching modes of TPS modulation, the equivalent circuit in Fig. 7.5 is used. It is possible to assess $i_L(t)$ and i_{pk} segment by segment within a switching period by using the principle of inductor volt-second balance piecewise for specified values of D_0, D_1 and D_2 [22, 23].

There are two issues with this approach. First off, it takes a lot of time and effort to manually derive the expressions of $i_L(t)$ and i_{pk} for all switching modes. Figure 7.6 provides a description of this procedure. Given that TPS modulation has a total of three modulation variables and 12 switching modes that take into account both power transfer directions, the resulting expression current stress i_{pk}, which has been published in [18, 22, 28], has a high degree of complexity. Second, the analysis's accuracy is compromised by the presumptions of a lossless component and a minimal magnetizing current.

7.2.2.2 Challenge in Stage III: Realization of TPS Modulation

The ideal values of the modulation variables acquired in Stage II will be saved in a lookup table in order to implement TPS modulation in real-world applications. This lookup table will be kept in the microcontroller's memory. When operating in real-time, the lookup table will be consulted to retrieve the optimal values of the modulation variables (D_0, D_1 and D_2) based on the present operating conditions (P and V_2) [23].

The data recorded is discrete due to the design of the lookup table as depicted in Fig. 7.7. So it's possible that (P, V_2) will fall somewhere between $(P, V_2)_i$ and $(P, V_2)_{i+1}$. In this case, either $(P, V_2)_i$ or $(P, V_2)_{i+1}$ will be thought of as approximations, even though their (D_0, D_1, D_2) aren't the best for that particular operating condition

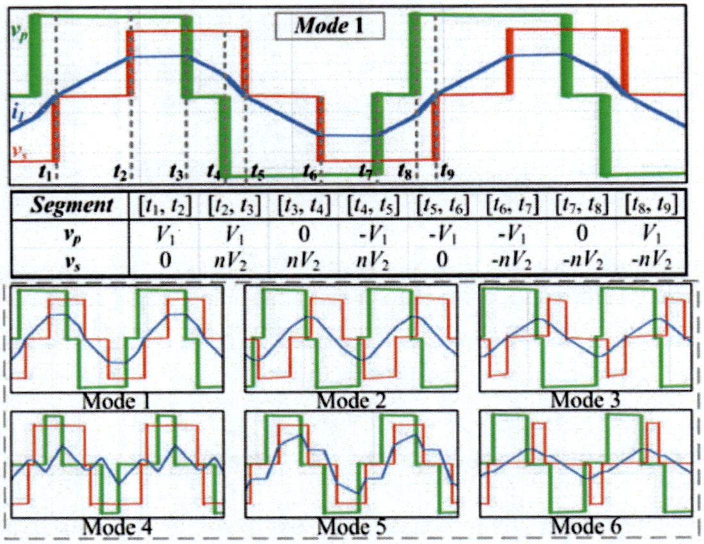

Fig. 7.6 Challenge in Stage I: complex segment-by-segment analysis of current stress under different operating modes of TPS modulation

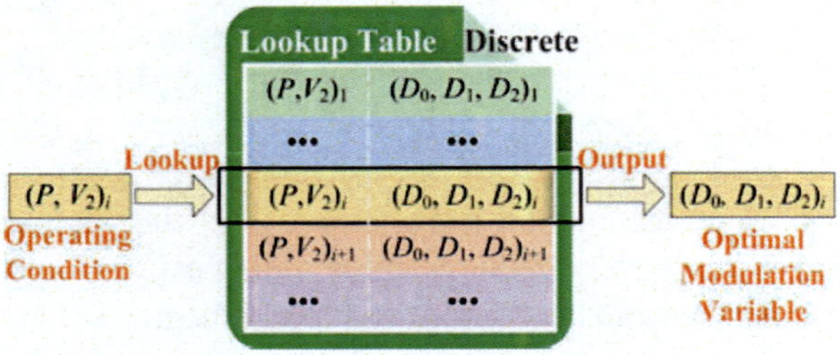

Fig. 7.7 Challenge II: discrete nature of lookup table for the realization of TPS modulation

[26]. Probabilities are that the derived modulation variables do not work at their best, decreasing accuracy.

In conclusion, there are certain issues with high complexity and low accuracy with the current optimization methodologies of TPS modulation for minimal current stress that need to be addressed.

7.3 The Proposed AI-Based TPS Modulation

In this chapter, an AI-based TPS modulation (AI-TPSM) optimization technique to reduce current stress is proposed to address the issues mentioned above. The AI-TPSM typically has three stages, each with a unique AI tool. Figure 7.8 details the full AI-TPMS optimization procedure.

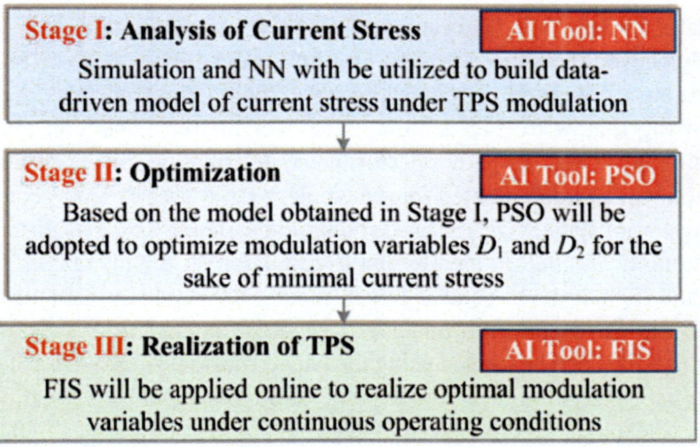

Fig. 7.8 Descriptions for the proposed AI-TPSM

7.3.1 *Stage I: NN-Based Analysis of Current Stress*

In this chapter, a data-driven model of current stress is realized via NN, addressing the difficulty of complex manual analysis, and attaining high-level automation in analysis. The specific steps of Stage I are listed in Fig. 7.9 and are shown as follows.

All operating specifications, such as the input voltage V_1, output voltage V_2, output power P, and switching frequency f_s, should be determined before Stage I can begin.

First, combinations of modulation variables (D_1, D_2) and operating conditions (P, V_2) are specified. $N_1 1$ number of P values and N_2 number of V_2 values are randomly chosen from the ranges [P_{min}, P_{max}] and [V_{2_min}, V_{2_max}], respectively, for the operating conditions P and V_2. Modulation variables D_1 and D_2 are both in [0,

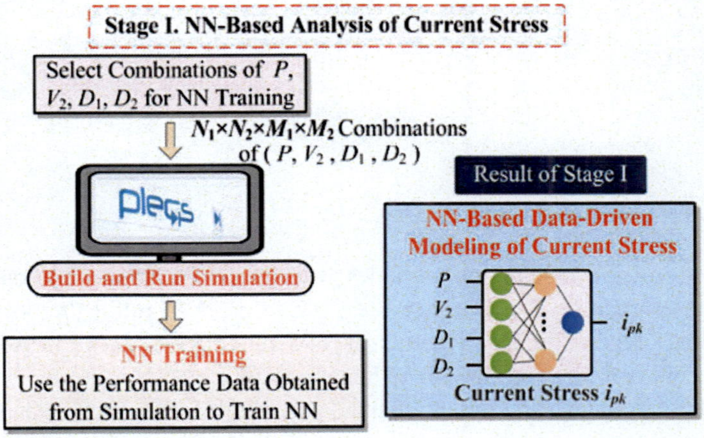

Fig. 7.9 Flowchart of Stage I: analysis of current stress with NN

1], as stated in Sect. 7.2. M_1 number of D_1 values and M_2 number of D_2 values are randomly chosen from the range [0, 1]. Thus, there are $N_1 \times N_2 \times M_1 \times M_2$ total combinations of operating circumstances and modulation variables.

Following that, a PLECS simulation model for the DAB converter with TPS modulation is created. To gather performance information for the current stress, this simulation will be implemented for the whole $N_1 \times N_2 \times M_1 \times M_2$ combinations of operating circumstances (P, V_2) and modulation variables (D_1, D_2). In the simulation, D_0 is automatically determined by the PI regulator for the given combination of P, V_2, D_1, and D_2 in order to control power transfer flow and maintain the stable output voltage of the DAB converter under TPS modulation [18].

After that, NN can be trained using the data on current stress performance that was obtained using the simulation model. The training set (70%), validating set (15%), and testing set (15%) of the total $N_1 \times N_2 \times M_1 \times M_2$ number of simulation data are used to train NN, choose NN structure, and test the trained NN on fresh and untested data points. Following training, NN can function as a data-driven model of current stress that can assess the performance of current stress under every conceivable set of operating conditions and modulation variables. It should be emphasized that peak current through inductor L (i_{pk}) is used to measure current stress.

7.3.2 *Stage II: Optimization with PSO Algorithm*

On the basis of the data-driven model created in Stage I, Stage II will determine the best modulation variables D_1, D_2 under various operating conditions P, V_2.

The optimization issue can be represented mathematically in the following way:

For the given operating conditions P and V_2, the goal is:

$$i_{pk}^* = \min_{D_1, D_2} i_{pk}(P, V_2, D_1, D_2) \tag{7.3}$$

Subject to:

$$0 \le D_1 \le 1 \tag{7.4}$$

$$0 \le D_2 \le 1 \tag{7.5}$$

D_0, one of the modulation variables, will be determined by the output of the PI regulator once D_1 and D_2 have been set, in order to fulfill the power transfer and voltage regulation requirements of TPS modulation. D_0 is therefore not an independent optimization variable to be taken into account.

The best modulation variables D_1 and D_2 are obtained by using the particle swarm optimization (PSO) algorithm to address this optimization problem. PSO algorithm is an evolutionary algorithm that searches for the best outcomes in the solution space by modeling the behavior of flocks of birds [29, 30]. The particle position's dimension is set to 2, consisting of two modulation variables, in order to solve the optimization problem in (7.3). (D_1, D_2). The values of D_1 and D_2 are represented by the particle's position X. Every iteration's location change is represented by the particle's velocity or V. The updates to each particle's position and velocity in the 2-dimensional space are given in (7.6) and (7.7), respectively.

$$V_i^{m+1} = \omega V_i^m + c_1 r_1 (P_{best\,i}^m - X_i^m) + c_2 r_2 (G_{best}^m - X_i^m) \tag{7.6}$$

$$X_i^{m+1} = X_i^m + V_i^m \tag{7.7}$$

If P_{best} is the personal best information and G_{best} is the global best information, m is the iteration number, c_1 and c_2 are the learning coefficients, ω is the inertia weight.

Figure 7.10 illustrates the full procedure of Stage II to identify the ideal modulation variables D_1 and D_2 under various operating situations. The first two are the operational conditions, P and V_2. After startup, each cycle evaluates the objective value i_{pk} for each particle separately. Additionally, each particle's position and velocity will be modified in accordance with (7.6) and (7.7). When the stopping requirement is satisfied, the discovered optimal $i_{pk}{}^*$ and the accompanying modulation variables $D_1{}^*$ and $D_2{}^*$ under the specified P, V_2 are then preserved. Once the best D_1 and D_2 have been identified for each combination of P and V_2, the process is repeated.

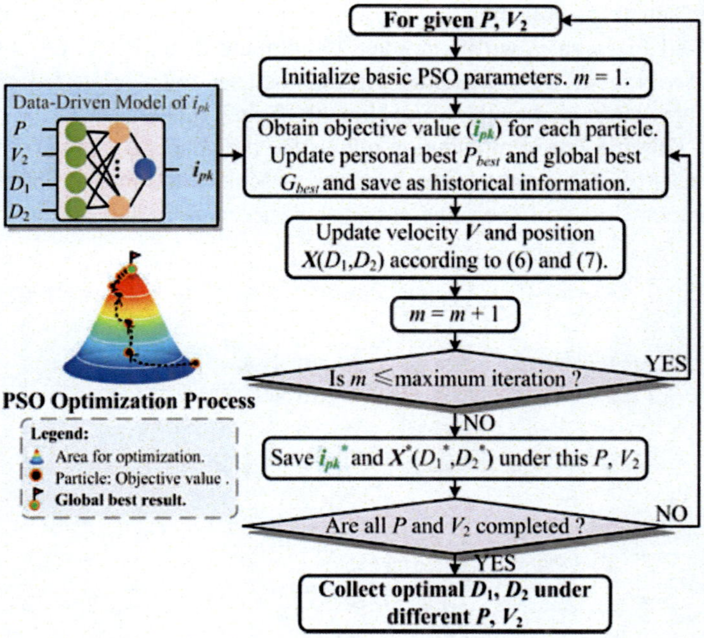

Fig. 7.10 Flowchart of Stage II: PSO for optimization

To achieve minimal current stress performance, the PSO algorithm in Stage II can obtain optimal D_1 and D_2 for the given combinations of P and V_2.

7.3.3 Stage III: Realization of TPS with FIS

The fuzzy inference system (FIS) is utilized to achieve continuous and accurate TPS modulation in real-time applications, avoiding the inaccurate optimum modulation variables generated by discrete lookup tables. Because of its excellent interpretability and higher generalization power, FIS is preferred over traditional interpolation techniques like cubic interpolation [31]. Figure 7.11 provides the FIS-based control diagram for DAB with TPS modulation as proposed in this chapter.

One of the three modulation variables (D_0, D_1, D_2) should be tuned by the PI regulator in order to satisfy the demands of power transfer and voltage regulation under TPS modulation. The phase shift D_0 between two full bridges is selected as the output of the PI module in the proposed FIS-based closed-loop control diagram for DAB under TPS modulation, with the error between the reference voltages $V_{2,ref}$ and V_2 acting as the module's input.

The FIS modulator chooses the settings for the duty cycles D_1 and D_2 inside two complete bridges. Figure 7.11 illustrates the FIS's basic operating principle for achieving online TPS modulation. The input membership functions of the FIS modulator will calculate the degree to which $V_{2,ref}$ and P belong to each of the linguistic sets (e.g., P is high, $V_{2,ref}$ is low). Then, with N being the total number of rules, all fuzzy rules will be applied to compute their O_1, … O_N outputs. The weighted average of all the outputs will then be used to calculate D_1 and D_2, with the weights determined by the firing strength of each rule [32].

Stage III's FIS-based modulation allows for the adaptive tuning of modulation variables D_1 and D_2 in online applications to achieve the lowest possible current stress under a variety of operating circumstances P and V_2. According to Fig. 7.12,

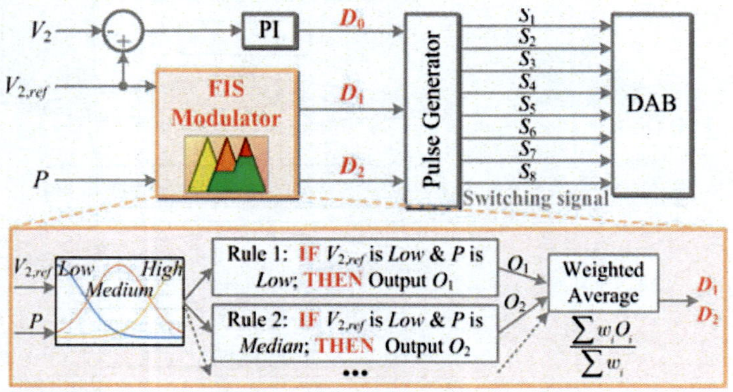

Fig. 7.11 Stage III: online realization of TPS with FIS

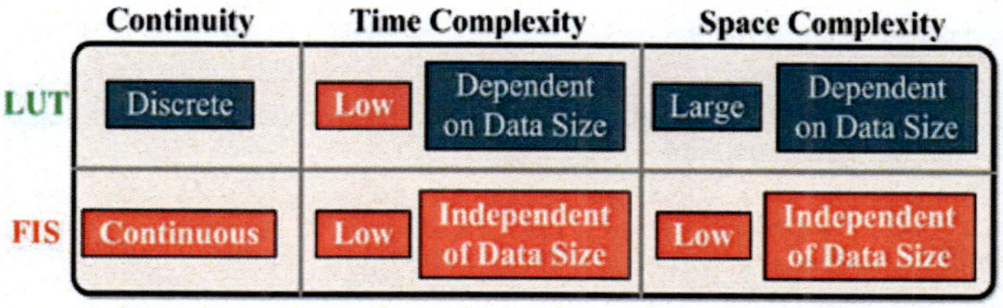

Fig. 7.12 Comparisons between LUT and the adopted FIS

FIS-based TPS modulation can achieve the best modulation variables D_1 and D_2 in the real-world circumstances of changing operating conditions P and V_2 with continuous values, as opposed to discrete lookup tables (LUT). Fast processing speeds are shared by LUT and FIS in terms of time complexity. LUT requires a lot of storage space, whereas FIS has a low space complexity. Additionally, the complexity of FIS is independent of the size of the data, whereas the complexity of LUT increases as the size of the data increases [33].

By training an NN-based data-driven model of current stress in Stage I of the proposed AI-TPSM, the laborious manual process of deriving current stress is reduced to a minimum. In Stage II, PSO is used to reduce the present stress for a set of operational conditions. In Stage III, a FIS-based control diagram for TPS modulation is proposed to get the best modulation variables for continuous operation in real-time applications.

7.4 Design Case of Applying the Proposed AI-TPSM

A current-stress-optimal TPS modulation scheme for the DAB converter is created using the proposed AI-TPSM technique described in Sect. 7.3. The following stage-by-stage illustration shows the design case.

In this design scenario, the output power P and output voltage V_2 of the DAB converter can range from 100 to 1000 W and 160 V to 230 V, respectively. The operating conditions and requirements of the DAB converter under TPS modulation are provided in Table 7.1. The choice of C2M0080120D is made for security purposes. The proposed AI-TPSM is still relevant even when additional power switches are taken into account.

Only P and V_2 are taken into account as changeable operating conditions in this chapter for the convenience of the presentation and computation. Incorporating the changing of V_1 into simulation and NN training (Stage I), optimization with PSO (Stage II), and FIS-based online realization will still allow AI-TPSM to be used (Stage III).

Table 7.1 Design specifications

Rated operating specifications			
P	1000 W	V_2	200 V
V_1	200 V	f_s	20 kHz
Power switches			
Switches	C2M0080120D, Cree	Dead time	500 ns
$R_{DS(on)}$	80 mΩ	V_{DSS}	1200 V
High-frequency transformer			
Inductor L	140 μH		
Core material	Iron based nanocrystalline alloy		
Modulation variables			
Duty cycle of full bridge 1 D_1	Duty cycle of full bridge 2 D_2		
Operating conditions			
Output power P	Output voltage V_2		
Ranges of modulation variables			
D_1	$D_{1\_min} = 0$; $D_{1\_max} = 1$		
D_2	$D_{2\_min} = 0$; $D_{2\_max} = 1$		
Ranges of operating conditions			
P	$P_{min} = 100$ W; $P_{max} = 1000$ W		
V_2	$V_{2\_min} = 160$ V; $V_{2\_max} = 230$ V		

7.4.1 Stage I: NN-Based Analysis of Current Stress

In Stage I of the proposed AI-TPSM technique, an automatically generated NN-based data-driven model of present stress is obtained by following the flowchart in Fig. 7.9. The following is a summary of the steps.

- To begin with, a total of 20 × 20 × 20 × 20 (160,000 in total) combinations of P, V_2, D_1, and D_2 are evenly chosen to adequately cover their ranges for the advantages of NN training.
- Next, using the PLECS program, the simulation of a DAB converter with TPS modulation is constructed. This process is repeated for all the possible combinations of P, V_2, D_1, and D_2. In every simulation, the current stress performance is obtained.
- All performance data generated by simulations can be divided into three sets: a training set (70%), a validating set (15%), and a testing set (15%). On the basis of training data, an NN-based data-driven model of current stress can be created, with P, V_2, D_1, and D_2 serving as inputs and the current stress indicator i_{pk} serving as the output. The neural network's structure is determined by the validation set's lowest error, as shown in Table 7.2 and Fig. 7.13. The trained NN exhibits greater accuracy on all datasets when compared to alternative regression techniques [34] as

response surface, Bayesian regression, and support vector regression (SVR). The average percentage deviations of the trained NN are substantially lower than those of other approaches, as shown in Table 7.3. In all the datasets under consideration, NN still provides satisfactory accuracy even in the worst-case circumstances.

Table 7.2 Configuration of NN and its optimizer

Selected NN	
Inputs	P, V_2, D_1, D_2
Output	Current stress i_{pk}
Hidden layers	Two layers, each of which has 128 and 32 neurons with ReLU activations
NN optimizer	
Data for NN training	70%, 112,000
Data for structure selection	15%, 24,000
Data for testing new data	15%, 24,000
Optimizer	Adaptive subgradient method [35]
Learning rate	0.001
Regularization coefficient	1e−5
Maximal epochs	10,000

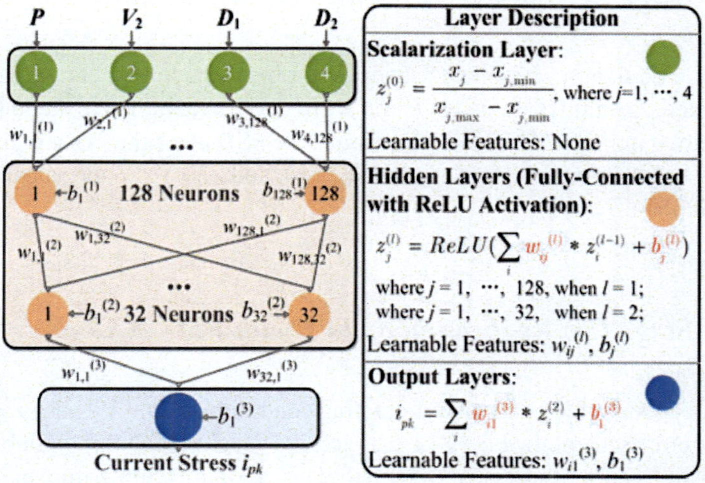

Fig. 7.13 Structure of the selected NN

Table 7.3 Accuracy of the trained NN

Percentage deviations	Training set		Validating set		Testing set	
	Average (%)	*Largest* (%)	*Average* (%)	*Largest* (%)	*Average* (%)	*Largest* (%)
Response surface	26.2	42.0	26.5	42.6	26.5	39.3
Bayesian regression	25.6	38.6	25.8	40.2	25.7	41.4
SVR	9.14	13.1	9.07	12.3	9.1	12.9
NN	**0.47**	**1.39**	**0.46**	**1.33**	**0.45**	**1.29**

Bold terms represents as to emphasize that '**NN**' provides satisfactory accuracy in all the datasets under consideration compared with other methods

Table 7.4 Configuration of PSO algorithm

Hyperparameter name	Hyperparameter value
Number of particles	20
Maximal number of iterations	100
Inertia weight ω	Linearly decrease from 0.9 to 0.4
Learning coefficients c_1, c_2	$c_1 = c_2 = 2.05$

7.4.2 Stage II: Optimization with PSO Algorithm

The best modulation variables D_1 and D_2 to obtain the lowest current stress under the selected operating parameters P and V_2 have been identified in Stage II using the PSO method in Fig. 7.10. Table 7.4 contains the PSO algorithm's settings. The ideal D_1 and D_2 for output voltages V_2 of 200, 160, and 230 V are graphically depicted in Fig. 7.14a–c.

7.4.3 Stage III: Realization of TPS with FIS

Under all conceivable continuous operating conditions, P and V_2, the proposed FIS-based control scheme in Fig. 7.11 can achieve ideal modulation variables D_1 and D_2 with little current stress in Stage III. This design example implements type-1 Takagi–Sugeno FIS [31], using gaussian input membership functions for P and V_2 and three linguistic sets (low, medium, and high). In Table 7.5, the configuration of FIS is enumerated. In Fig. 7.15a, b, the plots on the left display the FIS's outputs (D_1, D_2) in relation to the continuous inputs P and V_2, while the plots on the right display sample discrete values that were saved in a lookup table. Figure 7.15 illustrates how the proposed FIS-based TPS outperforms lookup tables by enabling real-time modulation in all P and V_2 scenarios.

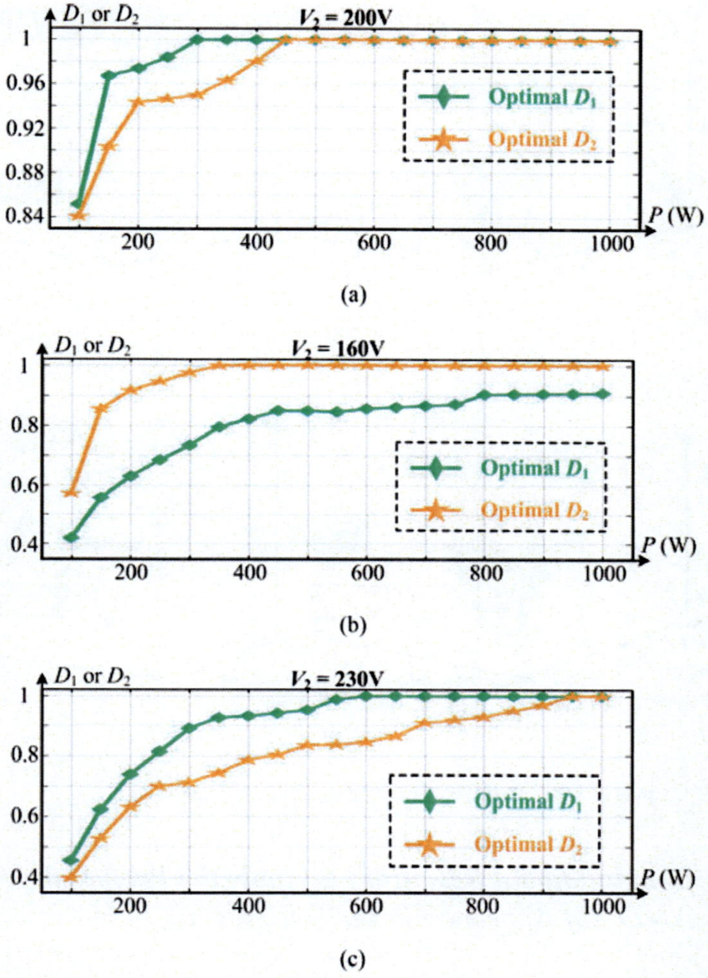

Fig. 7.14 Optimal D_1 and D_2 under different output voltage V_2 and different output power: **a** $V_2 = 200$ V; **b** $V_2 = 160$ V; **c** $V_2 = 230$ V

Table 7.5 Configuration of PSO algorithm

Inputs	P, V_2 of continuous values
Output	Optimal D_1, D_2
Type of FIS	Type-1 Takagi–Sugeno [31]
Type of input membership functions	Gaussian
Number of input membership functions	3
Number of rules	9

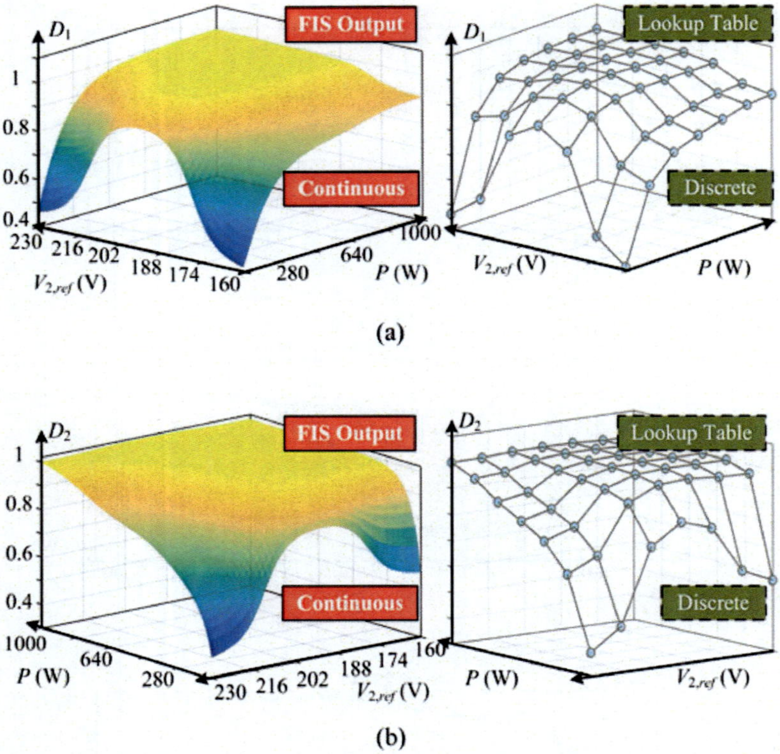

Fig. 7.15 Outputs of FIS with respect to the inputs P and V_2: **a** D_1; **b** D_2

7.4.4 Computational Resources to Apply the Proposed AI-TPSM Approach in the Design Case

The average CPU time of Stages I and II, as well as the average turnaround time and storage size of Stage III, are assessed and displayed in Table 7.6 to give an understanding of the computing resources needed to use AI-TPSM in the design case.

Stage I requires more CPU time and resources than Stage II does to run the simulation and train the NN. In Stage III, the turnaround time for deploying FIS

Table 7.6 Computational resources to apply AI-TPSM

	Platform	Performance
Stage I	Intel Xeon CPU E5-1630 @ 3.7 GHz, 16 GB RAM, Windows 10	*Average CPU Time*: 3 days and 21 h with four CPU cores
Stage II		*Average CPU Time*: 1.57 h
Stage III	Dspace 1202	*Average Turnaround Time*: 4.64 μs *Storage Size*: 3.4 *k*B

online in the control platform Dspace 1202 is only 4.64 μs, suggesting that FIS has a quick processing speed, and the storage size of FIS is just 3.4 kB, demonstrating that FIS has a simple spatial architecture. Comparatively, when a lookup table is used online, the storage space required might be substantial—up to several MB—which results in an undesirable turnaround time [33].

7.5 Experimental Verification

The proposed AI-TPSM approach has been tested in this part through experiments to provide the best TPS modulation with the least amount of current stress in the design case. The hardware platform for the hardware tests is depicted in Fig. 7.16, and design characteristics are provided in Table 7.1.

The experiments in this section are divided into the following sections: rated operation, operation at various P and V_2 levels, operation under different load and voltage step conditions, comparisons of SPS modulation, lookup-table-based TPS modulation, and AI-TPSM, and comparisons of the experimental and theoretical findings of the optimal modulation.

7.5.1 Rated Operating Waveforms

The rated operational waveforms of the DAB converter under the ideal TPS modulation with the least amount of current stress, which is designed using the proposed AI-TPSM technique, are shown in Fig. 7.17. In Fig. 7.1, the measured waveforms' notations and orientations are shown. Since D_1 and D_2 are both ideals under rated conditions 1, the waveforms v_p and v_s are only square waves.

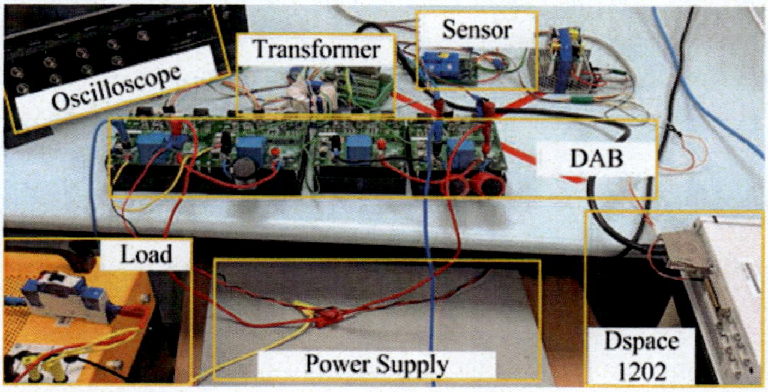

Fig. 7.16 Hardware platform in the experiments

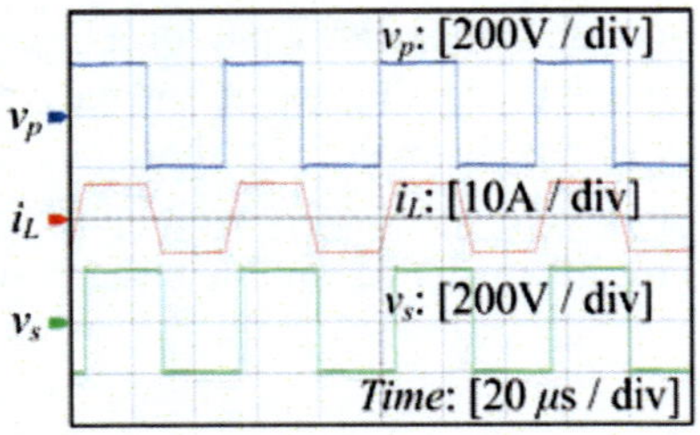

(a)

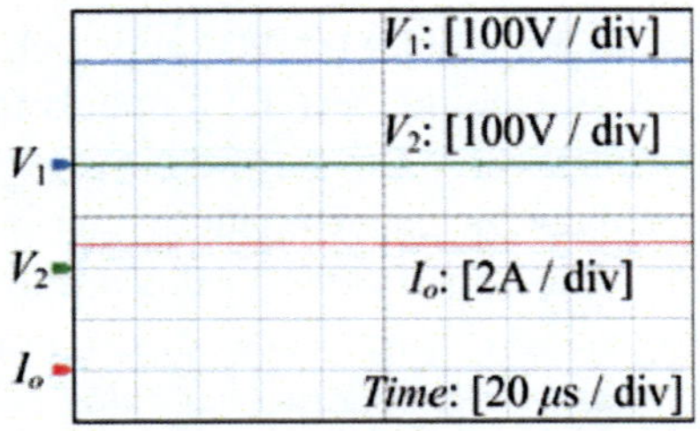

(b)

Fig. 7.17 Experimental waveforms under rated power of 1000W and rated output voltage of 200 V: **a** v_p, v_s, and i_L; **b** V_1, V_2 and I_o

7.5.2 *Operating Waveforms Under Different Output Power P and Output Voltage V_2*

In this section, experiments have been carried out with output voltages V_2 of 200, 160, and 230 V with output powers *P* of 100, 400, and 900 W.

Under V_2 of 200 V, the waveforms v_p, v_s, and i_L under 900W, 400W, and 100W are shown in Fig. 7.18; their corresponding operating modes are mode 1, mode 1, and mode 5.

When V_2 is 160 V, the waveforms v_p, v_s, and i_L under 900 W, 400 W, and 100 W, respectively, are presented in Fig. 7.19; their corresponding operating modes are mode 1, mode 1, and mode 4.

Given V_2 of 230 V, the waveforms v_p, v_s, and i_L under 900 W, 400 W, and 100 W are shown in Fig. 7.20; their corresponding operating modes are mode 1, mode 5, and mode 5.

7.5.3 *Transient Response Under Power and Voltage Step*

The tests below aim to validate the proposed FIS-based TPS modulation's real-time operation given the output power or voltage steps.

The output voltage V_2 steps from 200 V (1000 W) to 160 V (640 W) and back to 200 V when the load resistance is fixed at 40. (1000 W). The matching waveforms

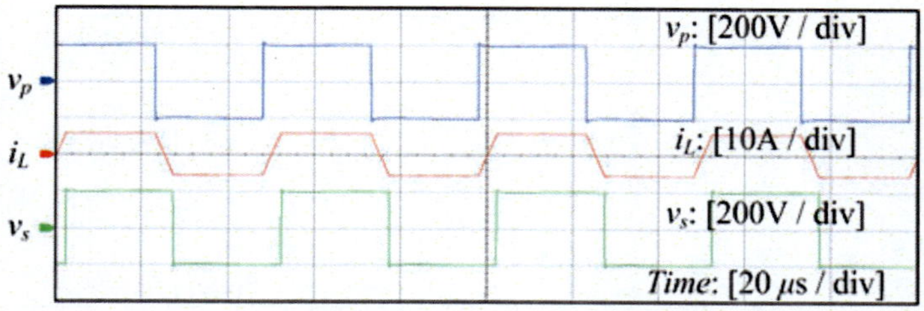

(a)

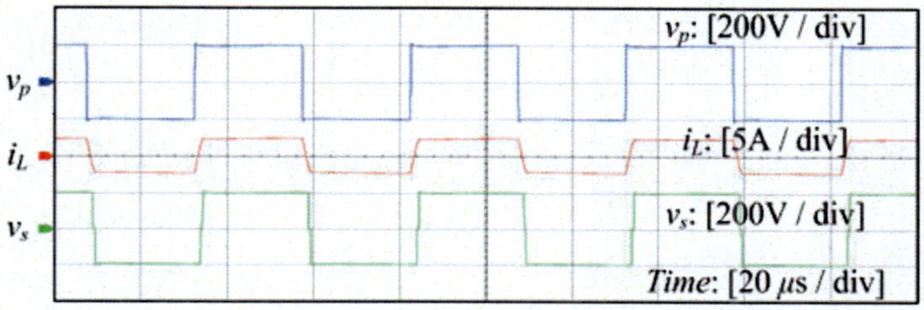

(b)

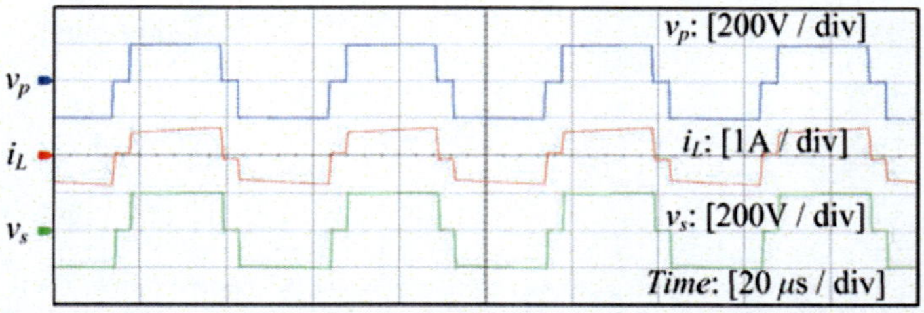

(c)

Fig. 7.18 Experimental waveforms under different output power when output voltage is 200 V: **a** $P = 900$ W; **b** $P = 400$ W; **c** $P = 100$ W

are shown in Fig. 7.21, with the modulation waveforms v_p, v_s, and i_L appearing at the bottom and the waveforms V_1, V_2, and I_o appearing at the top.

Second, the output voltage V_2 steps from 230 V (1000 W) to 200 V (756 W) and from 200 V (756 W) to 230 V when the load resistance is at 52.9 (1000 W). The waveforms are plotted in Fig. 7.22.

The transient responses of load steps under output voltages of 200 V, 160 V, and 230 V, respectively, are also shown in Figs. 7.23, 7.24, and 7.25.

According to the waveforms in Figs. 7.21 and 7.22, the proposed FIS-based TPS modulator can realize the real-time modifications of D_1 and D_2 to achieve the least amount of current stress when the operating conditions V_2 and P change. Additionally, output voltage V_2 can follow the necessary reference value when the output

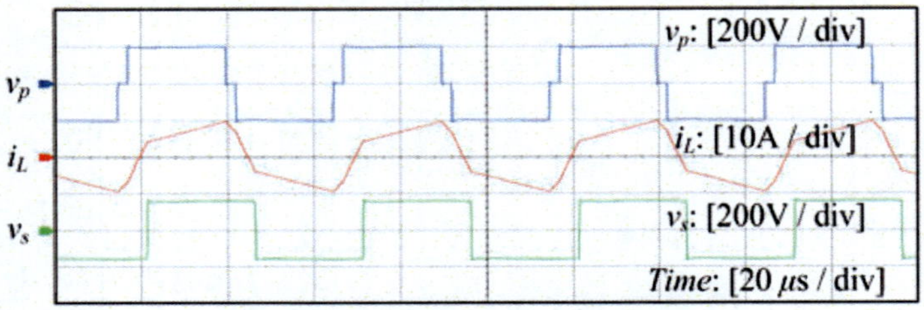

(a)

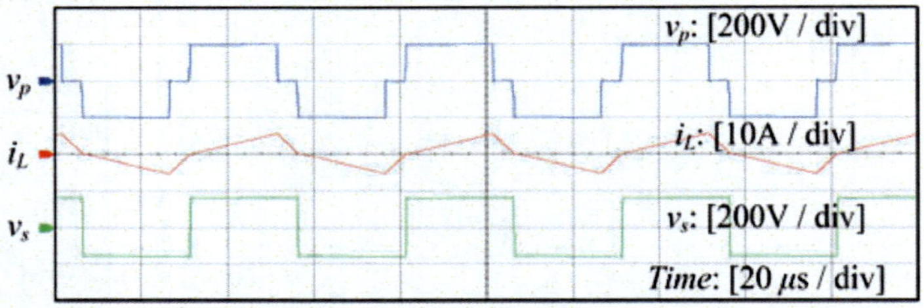

(b)

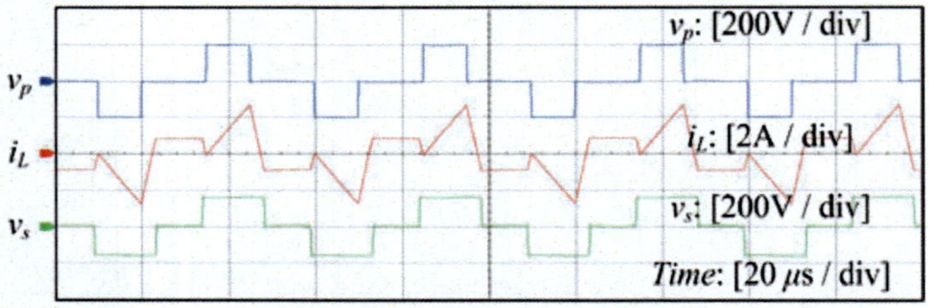

(c)

Fig. 7.19 Experimental waveforms under different output power when output voltage is 160 V: **a** $P = 900$ W; **b** $P = 400$ W; **c** $P = 100$ W

power P fluctuates, as shown in Figs. 7.23, 7.24 and 7.7.25, and D0, D1, and D2 have been optimized to their respective values.

In other words, the studies performed validated the proposed FIS-based TPS modulation's ability to operate online under various operating circumstances.

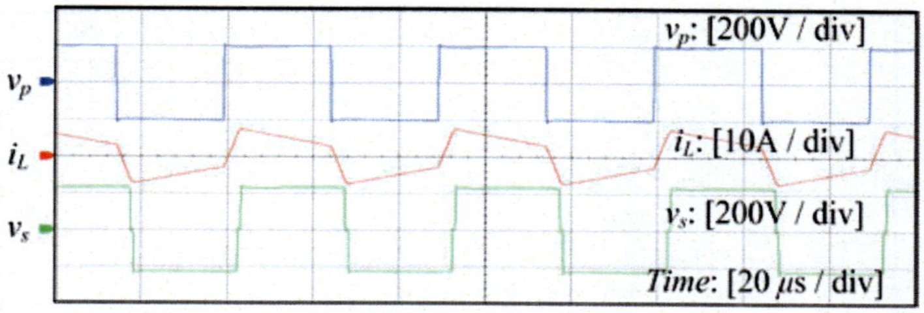

(a)

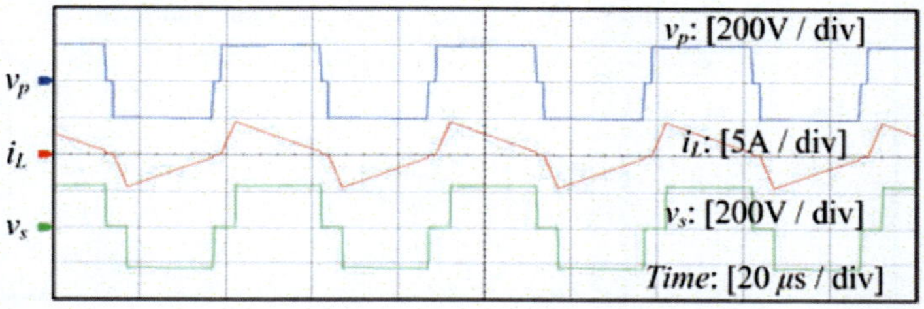

(b)

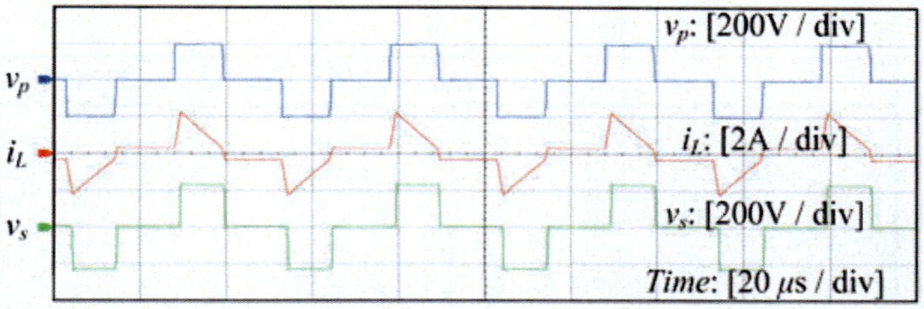

(c)

Fig. 7.20 Experimental waveforms under different output power when output voltage is 230 V: **a** $P = 900$ W; **b** $P = 400$ W; **c** $P = 100$ W

7.5.4 *Current Stress and Efficiency Performance of the Optimal TPS Modulation via the Proposed AI-TPSM*

The standard SPS modulation (SPSM) and the lookup-table-based TPS modulation (LUT-TPSM) are contrasted with the optimal TPS modulation using AI-TPSM in order to validate the suitable current stress and efficiency performance. When V_2 is 200 V, 160 V, or 230 V, respectively, Figs. 7.26, 7.27, and 7.28 show the current stress i_{pk} and efficiency η from 100 to 1000 W. In the experiments, efficiency is measured using the Teledyne LeCroy HDO8058A oscilloscope as the ratio of output power P_o to input power P_{in}.

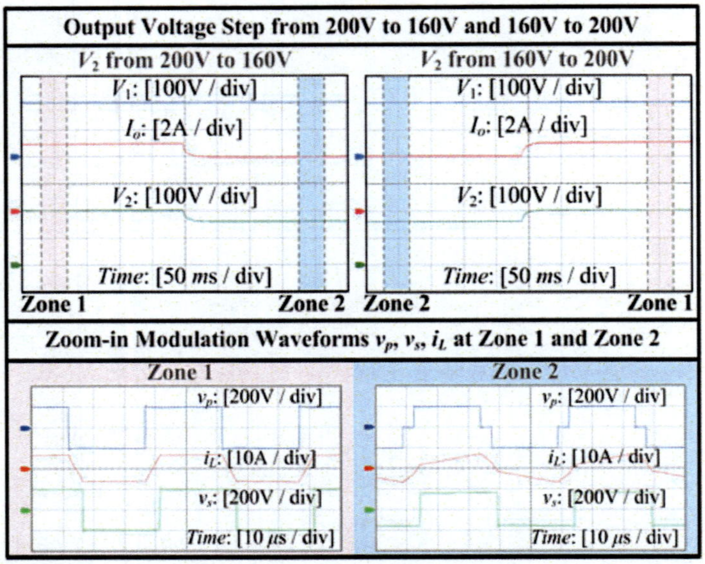

Fig. 7.21 Experimental waveforms given that V_2 steps from 200 to 160 V and V_2 steps from 160 to 200 V: V_1, V_2, and I_o during voltage step (top); zoom-in view of v_p, v_s, and i_L at Zone 1 and Zone 2 (bottom)

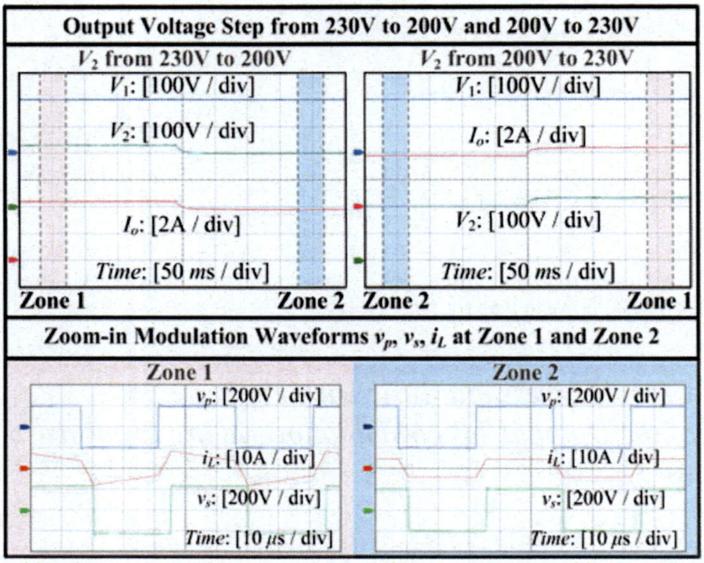

Fig. 7.22 Experimental waveforms given that V_2 steps from 230 to 200 V and V_2 steps from 200 to 230 V: V_1, V_2, and I_o during voltage step (top); zoom-in view of v_p, v_s, and i_L at Zone 1 and Zone 2 (bottom)

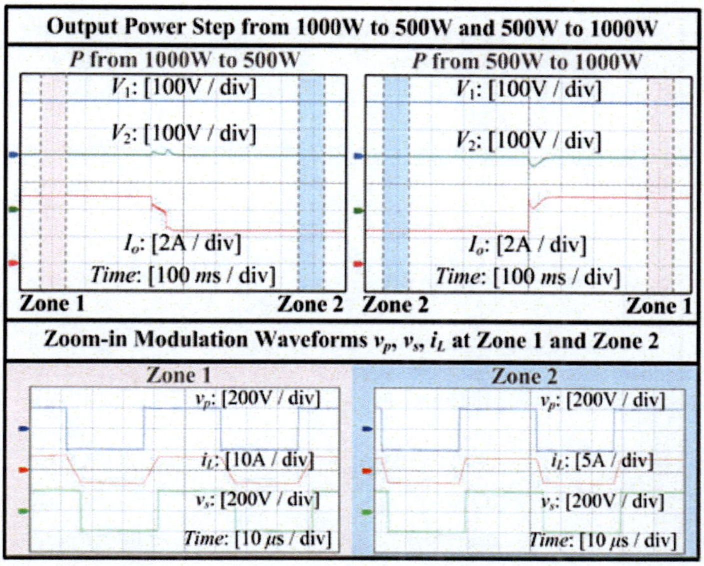

Fig. 7.23 Experimental waveforms given that P steps from 1000 to 500 W and from 500 to 1000 W under V_2 of 200 V: V_1, V_2, and I_o during power step (top); zoom-in view of v_p, v_s, and i_L at Zone 1 and Zone 2 (bottom)

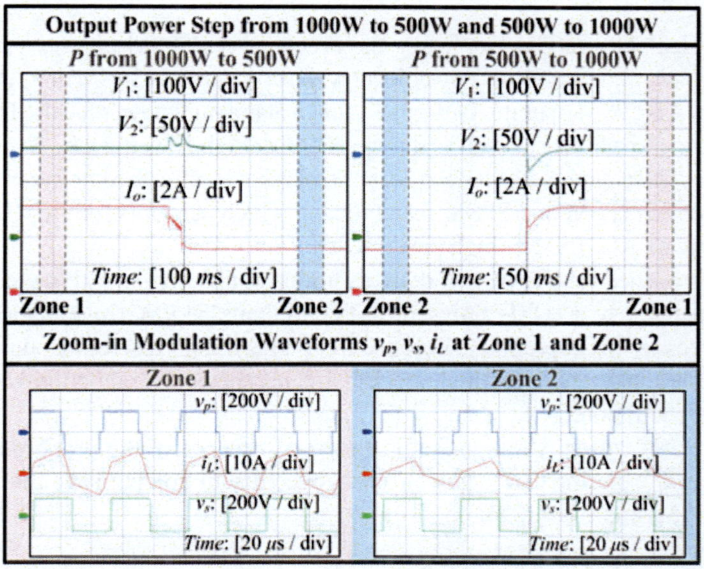

Fig. 7.24 Experimental waveforms given that P steps from 1000 to 500 W and from 500 to 1000 W under V_2 of 160 V: V_1, V_2, and I_o during power step (top); zoom-in view of v_p, v_s, and i_L at Zone 1 and Zone 2 (bottom)

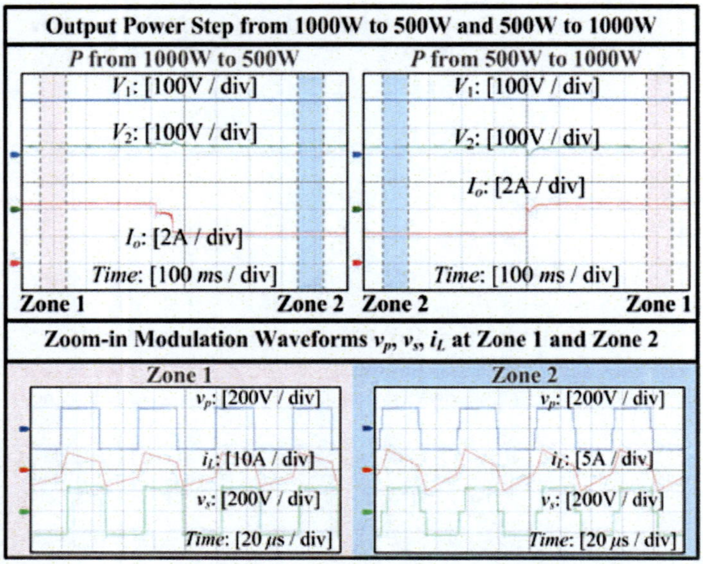

Fig. 7.25 Experimental waveforms given that P steps from 1000 to 500 W and from 500 to 1000 W under V_2 of 230 V: V_1, V_2, and I_o during power step (top); zoom-in view of v_p, v_s, and i_L at Zone 1 and Zone 2 (bottom)

The ideal TPS modulation using the proposed AI-TPSM approach has a much lower i_{pk} and higher η at low power level when compared to SPSM and AI-TPSM. In addition, AI-TPSM operates better in i_{pk} and η at medium power levels. The standard SPSM performs nearly as well as the AI-TPSM at high power levels. The optimal D_1 and D_2 in these circumstances are near 1, making the difference between the control signals of AI-TPSM and SPSM insignificant, which accounts for the relatively tiny advantage of AI-TPSM at high power levels.

The proposed AI-TPSM still delivers lower i_{pk} and higher η at low power levels in comparison to LUT-TPSM. The AI-continuous TPSM's modulation function contributes to this supremacy. Because LUT-TPSM uses discrete modulation, even though the lookup table in this model maintains the best modulation variables, the results are still poor.

The better current stress and efficiency performance of the proposed AI-TPSM approach is experimentally proven in this section in comparisons among SPSM, LUT-TPSM, and AI-TPSM.

7.5.5 Comparisons Between the Experimental and Theoretical Results of the Optimal Modulation

The proposed AI-TPSM approach's great accuracy and optimality are confirmed through a comparison between its experimental and theoretically ideal results.

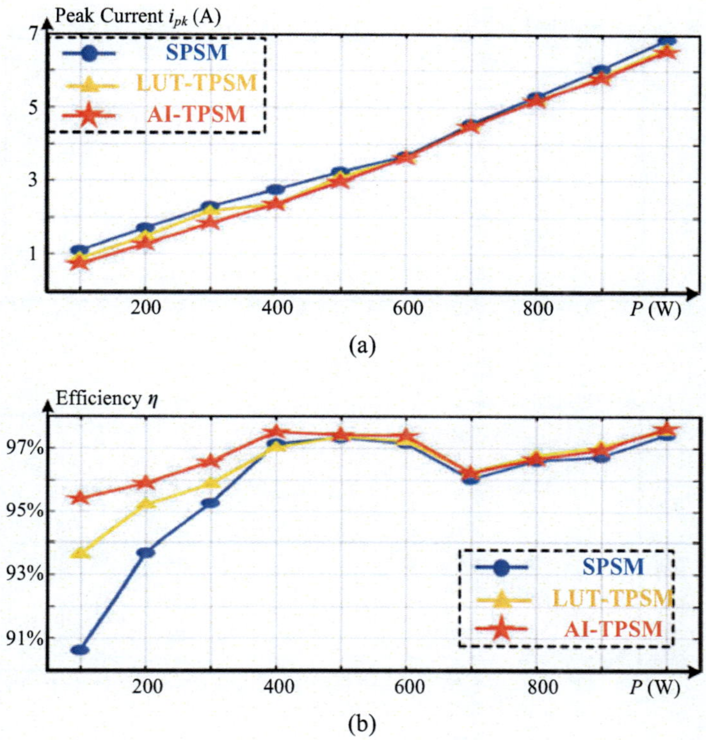

Fig. 7.26 Current stress i_{pk} and efficiency η performance of SPSM, LUT-TPSM and the optimal TPS modulation via AI-TPSM when output voltage V_2 is 200 V: **a** current stress i_{pk} performance; **b** efficiency η performance

When the output voltage is 200 V, 160 V, or 230 V, respectively, the average variances between the practical and theoretical findings are 2.96%, 3.51%, and 3.37%, as shown in Fig. 7.29. The highest deviation error is merely 5.21% in the worst-case scenario. Thus, the negligible discrepancies between the experimental and theoretically ideal findings verify the proposed AI-excellent TPSM's accuracy and demonstrate the accuracy of the experimental peak current data.

In other words, the design scenario is thoroughly verified with the hardware experimental results stated above, and the proposed AI-TPSM approach for optimal TPS modulation with minimum current stress is validated.

7.6 Conclusion

This chapter proposes an AI-TPSM, or artificial intelligence-based triple phase shift modulation optimization technique, for the dual active bridge converter that can automatically achieve reduced current stress. The analysis process uses neural networks,

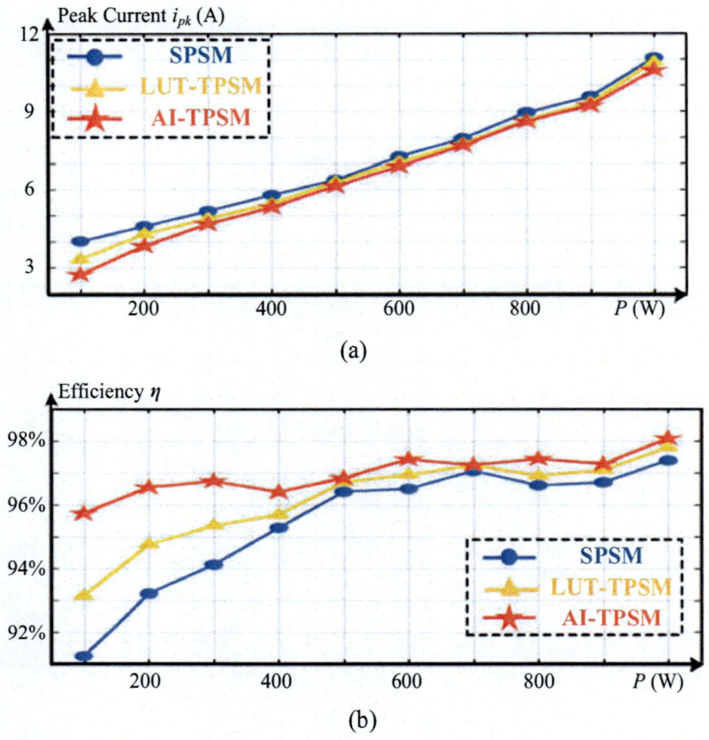

Fig. 7.27 Current stress i_{pk} and efficiency η performance of SPSM, LUT-TPSM, and the optimal TPS modulation via AI-TPSM when output voltage V_2 is 160 V: **a** current stress i_{pk} performance; **b** efficiency η performance

the optimization process uses evolutionary algorithms, and the realization process uses fuzzy inference systems. Generally speaking, AI-TPSM can be separated into three stages, each step containing one AI tool. First, a neural network is trained with simulations to learn the links between the variables (operation conditions and modulation variables) and the current stress in order to replace the conventionally laborious and unreliable analytical method. Second, an evolutionary approach called particle swarm optimization is used to identify the best outcomes for modulation that can reduce current stress. In order to offer continuous modulation, the best modulation results under various operating situations are stored using a fuzzy inference method. The proposed AI-TPSM has a high level of automation, which can lessen the workload for engineers and increase accuracy. Finally, the effectiveness of the AI-TPSM has been verified with a 1 kW prototype of the DAB converter.

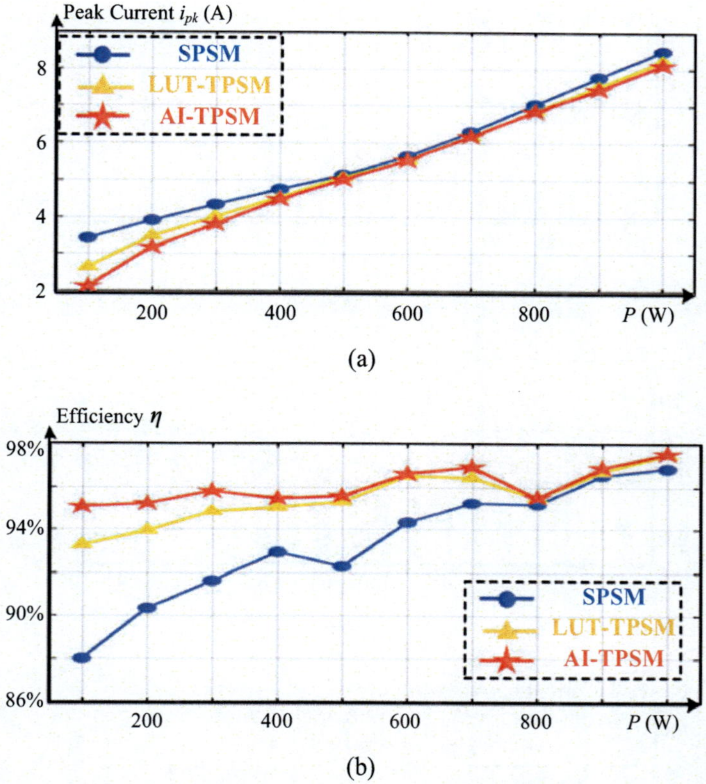

Fig. 7.28 Current stress i_{pk} and efficiency η performance of SPSM, LUT-TPSM, and the optimal TPS modulation via AI-TPSM when output voltage V_2 is 230 V: **a** current stress i_{pk} performance; **b** efficiency η performance

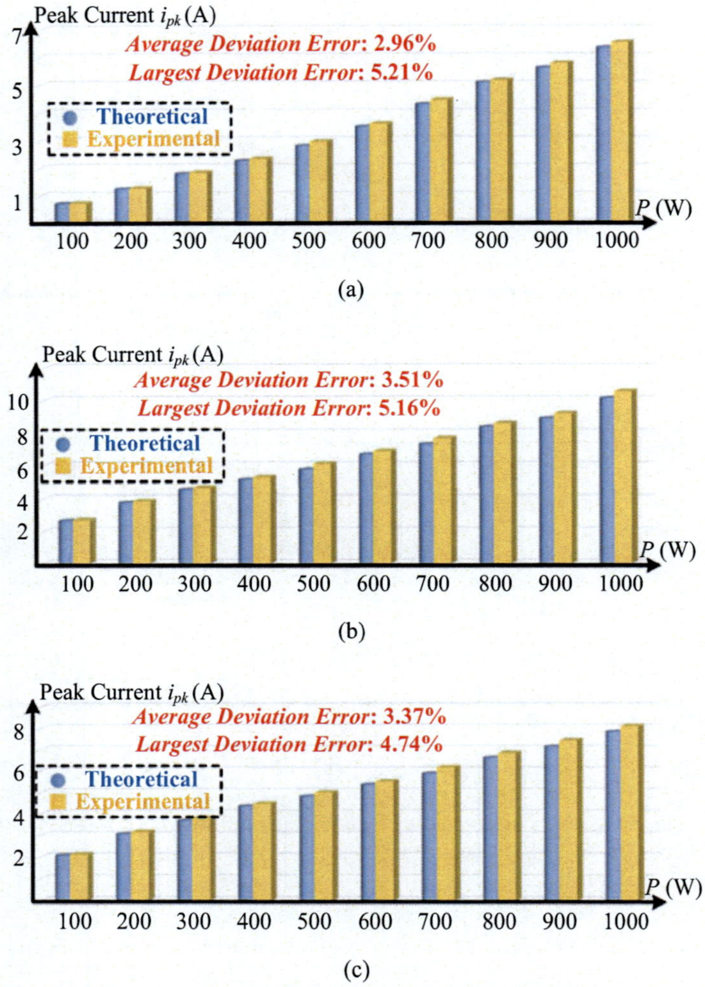

Fig. 7.29 Comparisons between the experimental results and the theoretically optimal results: **a** $V_2 = 200$ V; **b** $V_2 = 160$ V; **c** $V_2 = 230$ V

References

1. M. N. Kheraluwala, R. W. Gascoigne, D. M. Divan, and E. D. Baumann, 'Performance characterization of a high-power dual active bridge DC-to-DC converter', *IEEE Trans. Ind. Appl.*, vol. 28, no. 6, pp. 1294–1301, Dec. 1992.
2. J. Huang, X. Zhang, and C. Wen, 'Two-Stage Parameter Design Methodology of a Generalized Resonant DC Transformer in Hybrid AC/DC Microgrid With Optimum Active Power Transmission', *IEEE J. Emerg. Sel. Top. Power Electron.*, vol. 8, no. 3, pp. 2313–2325, Sep. 2020.
3. F. Lin, X. Zhang, X. Li, H. Ma, and C. Cai, 'Design of Symmetrical CLLC Resonant DC Transformer Considering Voltage Transfer Ratio and Cascaded System Stability', *IEEE Trans. Ind. Electron.*, pp. 1–1, 2021.

4. F. Lin, X. Zhang, X. Li, C. Liu, and H. Ma, 'Parameter Design for Symmetrical CLLC-Type DC Transformer Considering Cascaded System Stability and Power Efficiency', *IEEE J. Emerg. Sel. Top. Power Electron.*, pp. 1–1, 2020.
5. C. Sun, X. Zhang, and X. Cai, 'A Step-Up Nonisolated Modular Multilevel DC–DC Converter With Self-Voltage Balancing and Soft Switching', *IEEE Trans. Power Electron.*, vol. 35, no. 12, pp. 13017–13030, Dec. 2020.
6. A. K. Jain and R. Ayyanar, 'Pwm control of dual active bridge: Comprehensive analysis and experimental verification', *IEEE Trans. Power Electron.*, vol. 26, no. 4, pp. 1215–1227, Apr. 2011.
7. F. Krismer and J. W. Kolar, 'Accurate Small-Signal Model for the Digital Control of an Automotive Bidirectional Dual Active Bridge', *IEEE Trans. Power Electron.*, vol. 24, no. 12, pp. 2756–2768, Dec. 2009.
8. B. Zhao, Q. Yu, and W. Sun, 'Extended-Phase-Shift Control of Isolated Bidirectional DC–DC Converter for Power Distribution in Microgrid', *IEEE Trans. Power Electron.*, vol. 27, no. 11, pp. 4667–4680, Nov. 2012.
9. B. Liu, P. Davari, and F. Blaabjerg, 'An Optimized Hybrid Modulation Scheme for Reducing Conduction Losses in Dual Active Bridge Converters', *IEEE J. Emerg. Sel. Top. Power Electron.*, vol. 9, no. 1, pp. 921–936, Feb. 2021.
10. B. Zhao, Q. Song, and W. Liu, 'Power Characterization of Isolated Bidirectional Dual-Active-Bridge DC–DC Converter With Dual-Phase-Shift Control', *IEEE Trans. Power Electron.*, vol. 27, no. 9, pp. 4172–4176, Sep. 2012.
11. Hua Bai and C. Mi, 'Eliminate Reactive Power and Increase System Efficiency of Isolated Bidirectional Dual-Active-Bridge DC–DC Converters Using Novel Dual-Phase-Shift Control', *IEEE Trans. Power Electron.*, vol. 23, no. 6, pp. 2905–2914, Nov. 2008.
12. Q. Gu, L. Yuan, J. Nie, J. Sun, and Z. Zhao, 'Current Stress Minimization of Dual-Active-Bridge DC–DC Converter Within the Whole Operating Range', *IEEE J. Emerg. Sel. Top. Power Electron.*, vol. 7, no. 1, pp. 129–142, Mar. 2019.
13. K. Wu, C. W. de Silva, and W. G. Dunford, 'Stability Analysis of Isolated Bidirectional Dual Active Full-Bridge DC–DC Converter With Triple Phase-Shift Control', *IEEE Trans. Power Electron.*, vol. 27, no. 4, pp. 2007–2017, Apr. 2012.
14. S. Wang, C. Li, K. Wang, Z. Zheng, and Y. Li, 'Loss Imbalance and Transient DC-Bias Mitigation in Dual-Active-Bridge DC/DC Converters', *IEEE J. Emerg. Sel. Top. Power Electron.*, vol. 9, no. 2, pp. 1399–1409, Apr. 2021.
15. P. Wang, X. Chen, C. Tong, P. Jia, and C. Wen, 'Large- and Small-Signal Average-Value Modeling of Dual-Active-Bridge DC-DC Converter with Triple-Phase-Shift Control', *IEEE Trans. Power Electron.*, pp. 1–1, 2021.
16. A. K. Bhattacharjee and I. Batarseh, 'Optimum Hybrid Modulation for Improvement of Efficiency Over Wide Operating Range for Triple-Phase-Shift Dual-Active-Bridge Converter', *IEEE Trans. Power Electron.*, vol. 35, no. 5, pp. 4804–4818, May 2020.
17. B. Zhao, Q. Song, and W. Liu, 'Efficiency Characterization and Optimization of Isolated Bidirectional DC–DC Converter Based on Dual-Phase-Shift Control for DC Distribution Application', *IEEE Trans. Power Electron.*, vol. 28, no. 4, pp. 1711–1727, Apr. 2013.
18. A. Tong, L. Hang, G. Li, X. Jiang, and S. Gao, 'Modeling and Analysis of a Dual-Active-Bridge-Isolated Bidirectional DC/DC Converter to Minimize RMS Current With Whole Operating Range', *IEEE Trans. Power Electron.*, vol. 33, no. 6, pp. 5302–5316, Jun. 2018.
19. N. Hou, W. Song, and M. Wu, 'Minimum-Current-Stress Scheme of Dual Active Bridge DC-DC Converter With Unified-phase-shift Control', *IEEE Trans. Power Electron.*, pp. 1–1, 2016.
20. L. Zhu, A. R. Taylor, G. Liu, and K. Bai, 'A Multiple-Phase-Shift Control for a SiC-Based EV Charger to Optimize the Light-Load Efficiency, Current Stress, and Power Quality', *IEEE J. Emerg. Sel. Top. Power Electron.*, vol. 6, no. 4, pp. 2262–2272, Dec. 2018.
21. Yanhui Xie, Jing Sun, and J. S. Freudenberg, 'Power Flow Characterization of a Bidirectional Galvanically Isolated High-Power DC/DC Converter Over a Wide Operating Range', *IEEE Trans. Power Electron.*, vol. 25, no. 1, pp. 54–66, Jan. 2010.

22. O. M. Hebala, A. A. Aboushady, K. H. Ahmed, and I. Abdelsalam, 'Generic Closed-Loop Controller for Power Regulation in Dual Active Bridge DC–DC Converter With Current Stress Minimization', *IEEE Trans. Ind. Electron.*, vol. 66, no. 6, pp. 4468–4478, Jun. 2019.
23. J. Huang, Y. Wang, Z. Li, and W. Lei, 'Unified Triple-Phase-Shift Control to Minimize Current Stress and Achieve Full Soft-Switching of Isolated Bidirectional DC–DC Converter', *IEEE Trans. Ind. Electron.*, vol. 63, no. 7, pp. 4169–4179, Jul. 2016.
24. F. Krismer and J. W. Kolar, 'Closed Form Solution for Minimum Conduction Loss Modulation of DAB Converters', *IEEE Trans. Power Electron.*, vol. 27, no. 1, pp. 174–188, Jan. 2012.
25. F. Krismer and J. W. Kolar, 'Efficiency-Optimized High-Current Dual Active Bridge Converter for Automotive Applications', *IEEE Trans. Ind. Electron.*, vol. 59, no. 7, pp. 2745–2760, Jul. 2012.
26. Y. Tang *et al.*, 'Reinforcement Learning Based Efficiency Optimization Scheme for the DAB DC-DC Converter with Triple-Phase-Shift Modulation', *IEEE Trans. Ind. Electron.*, pp. 1–1, 2020.
27. Y. A. Harrye, K. H. Ahmed, and A. A. Aboushady, 'Reactive power minimization of dual active bridge DC/DC converter with triple phase shift control using neural network', in *2014 International Conference on Renewable Energy Research and Application (ICRERA)*, Milwaukee, WI, USA, Oct. 2014, pp. 566–571.
28. S. S. Muthuraj, V. K. Kanakesh, P. Das, and S. K. Panda, 'Triple Phase Shift Control of an LLL Tank Based Bidirectional Dual Active Bridge Converter', *IEEE Trans. Power Electron.*, vol. 32, no. 10, pp. 8035–8053, Oct. 2017.
29. F. Lin, X. G. Zhang, and X. Li, 'Design Methodology for Symmetric CLLC Resonant DC Transformer Considering Voltage Conversion Ratio, System Stability and Efficiency', *IEEE Trans. Power Electron.*, pp. 1–1, 2021.
30. X. Li, K. Mao, F. Lin, and X. Zhang, 'Particle swarm optimization with state-based adaptive velocity limit strategy', *Neurocomputing*, vol. 447, pp. 64–79, Aug. 2021.
31. S. Coupland and R. John, 'Geometric Type-1 and Type-2 Fuzzy Logic Systems', *IEEE Trans. Fuzzy Syst.*, vol. 15, no. 1, pp. 3–15, Feb. 2007.
32. S. Zhao, F. Blaabjerg, and H. Wang, 'An Overview of Artificial Intelligence Applications for Power Electronics', *IEEE Trans. Power Electron.*, vol. 36, no. 4, pp. 4633–4658, Apr. 2021.
33. Y. Tang *et al.*, 'Artificial Intelligence-Aided Minimum Reactive Power Control for the DAB Converter Based on Harmonic Analysis Method', *IEEE Trans. Power Electron.*, pp. 1–1, 2021.
34. C. C. M. Chen, H. Schwender, J. Keith, R. Nunkesser, K. Mengersen, and P. Macrossan, 'Methods for Identifying SNP Interactions: A Review on Variations of Logic Regression, Random Forest and Bayesian Logistic Regression', *IEEE/ACM Trans. Comput. Biol. Bioinform.*, vol. 8, no. 6, pp. 1580–1591, Nov. 2011.
35. J. Duchi, E. Hazan, and Y. Singer, 'Adaptive Subgradient Methods for Online Learning and Stochastic Optimization', p. 39.

Chapter 8
The Proposed Optimal Design and Heterogeneous Integration of DC/AC Inverter for Electric Vehicle

8.1 Introduction

The power control unit (PCU), which links the power flow between the battery and motor as depicted in Fig. 8.1a [1], is the brain of an electric vehicle (EV). The crucial component of the PCU is the inverter, as shown in Fig. 8.1b. Due to the maximum operating temperature of Si power devices being limited to 120–150 °C, an extra 85 °C liquid cooling system is used for the Si-based inverter [2, 3] in addition to the 105 °C liquid cooling system for engines and motors. The next-generation SiC power device can operate beyond 200 °C with an extremely quick switching speed thanks to the high melting point, high thermal conductivity, high breakdown field, and high electron velocity [4, 5]. As a result, by using the SiC device, the additional 85 °C liquid cooling system can be replaced with air cooling, increasing the power density of the drivetrain. Additionally, the SiC device can be used to accomplish the increased switching frequency in EV inverters, which will result in smaller capacitor sizes and a much more compact PCU [6]. To meet the PCU's ever-increasing needs for lightness and compactness, the air-cooling SiC inverter is a promising EV contender.

The SiC-based inverters have been discussed in certain publications employing specialized power module packaging, the best passive component integration, and cutting-edge thermal management.

- **Concerning the power packaging**, High-temperature and high-power packaging are emphasized in order to fully take use of SiC MOSFETs' improved capabilities [7, 8]. High-temperature materials are used for die attachments, encapsulants, and casings to support the potential high junction temperature of SiC devices, increasing the operating temperature of the power module and decreasing the size of the heat sink [9, 10]. Additionally, [11–13] highlights the ideal chip architecture in a multi-chip power module in order to balance the parasitics of parallel power loops and minimize the equivalent packaging parasitics. In addition, certain unique packaging structures are emphasized, such as split-out bridge [12], double-end source [14], all planar [15], hybrid integration [16], flip-chip bonding [17], 3D stacked [18, 19], etc., to shorten the power loop and reduce parasitics.

X. Zhang et al., *Automated Design of Electrical Converters with Advanced AI Algorithms*,
https://doi.org/10.1007/978-981-99-0459-4_8

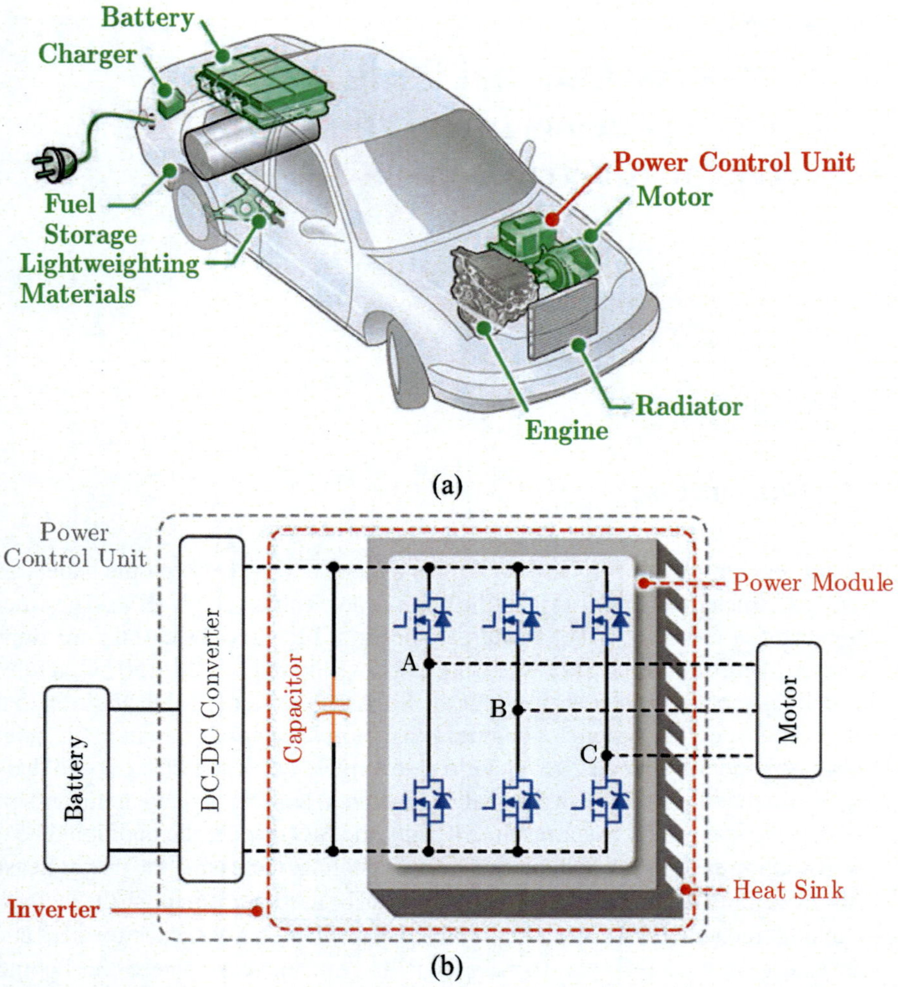

Fig. 8.1 Inverter-dominated power control unit for electric vehicle. **a** Configuration of electric vehicle and **b** schematic of power control unit

- **Concerning the dc-link capacitor**, The capacitor should be optimized taking into account material, ripple, and cost considerations in order to cost-effectively provide the dc-bus with a sufficient energy buffer [20, 21]. In [22–24], the electrolytic or film capacitors are evaluated in comparison to EVs and photovoltaic (PV) applications. Some sophisticated active capacitors that combine conventional capacitors and DC/DC converters are proposed [25–27] to increase the ripple current capabilities. To increase the capacitor's ability to handle ripple current and increase its power density, a stacked switching capacitor has been proposed [28]. Additionally, because SiC MOSFETs switch quickly, the EMI problem in an inverter is handled by employing the capacitor's optimal displacement, which is solved using evolutionary algorithms [29, 30].

- **Concerning the thermal management**, The SiC device's size is lower than its Si cousin, but because of the high switching frequency's power loss and resulting high heat flux, it presents a thermal management difficulty for the air-cooled SiC inverter. To increase power dissipation and decrease thermal resistance, some novel designs have been created, such as cold spray low-temperature soldering [31], integrated baseplate with Pin–Fin or Power-Shower technologies [13], 3D printed heat sinks [32], nanoscale thermal interface materials, and optimized heat sink structures [33, 34]. Additionally, active thermal management is offered from the perspective of the digital controller to online control the power losses of power devices, utilizing an adaptive switching frequency or an adjustable dc-link voltage [35].

The inverter's power module, dc-link capacitor, and heat sink are all crucial parts. For the installation of EVs, they share the most weight and volume of an inverter. They are uniquely created and developed in many references. These parts ought to be co-designed with an eye toward creating the best air-cooled SiC inverter possible. The air-cooling SiC inverter has no design approach, though. Furthermore, rather than using a multi-physics approach, the existing optimal designs for these crucial components concentrate on a single field. A multi-physics and multi-dimension perspective is not accessible for the best systemic designs of these components.

In this chapter, a systemic design methodology for an air-cooling SiC inverter is provided, aiming at the light and compact EV inverter without the liquid cooling system. From a multi-physics and multi-dimension perspective, detailed optimal design methodologies for the customized power module, dc-link capacitor, and heat sink are developed. This chapter's remaining sections are arranged as follows. The current condition and prospects of the air-cooling SiC inverter are discussed in Sect. 8.2. In Sect. 8.3, a tactical methodology for the SiC inverter is proposed. Section 8.4 presents specific optimal design strategies for the power module, dc-link capacitor, and heat sink. In Section V, prototypes and experiments showcasing the customized power module and integrated SiC inverter are presented. In Sect. 8.5, certain conclusions are drawn.

8.2 State-of-the-Art and Main Barriers of Air-Cooling SiC Inverter

The current situation of the EV inverter is briefly examined in this Section. Strategically, the use of an air-cooling SiC inverter is advised. Additionally, the key obstacles to an air-cooled SiC inverter with high power density are proposed.

8.2.1 State-of-the-Art of EV Inverter

The state-of-the-art and trend of EV inverters were shown in Fig. 8.2. As can be seen, Si inverters with liquid cooling control the market. The liquid-cooling SiC inverters, however, are taking the place of the Si inverter. Additionally, the next-generation PCU of the EV powertrain will feature air-cooling SiC inverters to meet the continuously rising demands of EVs for light and compactness.

With regard to the liquid cooling Si inverter, the EV has an additional 85 °C liquid cooling cycling system added in addition to the 105 °C cooling system. The additional liquid cooling system will inevitably decrease the power density, raise the price, and deteriorate the powertrain's reliability [36]. Utilizing cutting-edge power electronic components with high-temperature capabilities, such as SiC MOSFETs, the additional 85 °C liquid cooling can be substituted by straightforward air cooling, which can significantly outweigh the drawbacks of conventional liquid cooling PCU.

Figure 8.3a shows the thermal flux implementation ranges managed by various cooling systems. As can be observed, the forced air cooling system can handle heat flux of up to 200 W/cm^2. Figure 8.3b provides a summary of the power electronic device's thermal flux. The thermal flux of SiC MOSFET is determined to be approximately 120 W/cm^2, compared to Si IGBT's 10 W/cm^2. For the next-generation SiC-based EV inverter, air cooling might be the best option given its simplicity, low cost, high dependability, and small size. As a result, the trend in EV powertrain is the air-cooling SiC inverter.

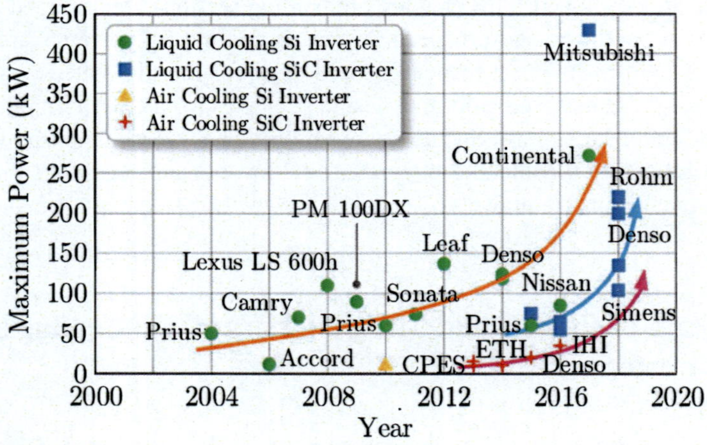

Fig. 8.2 State-of-the-art and trend of EV inverter

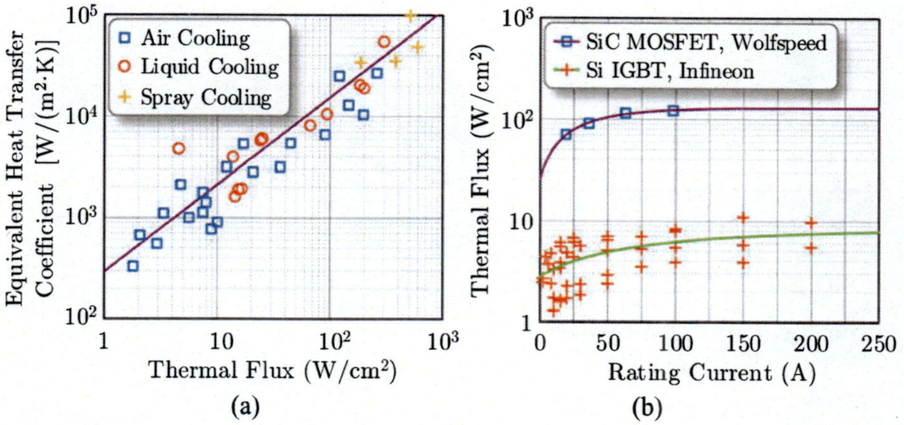

Fig. 8.3 Thermal flux of **a** coolant system capability and **b** power electronic chip

8.2.2 *Challenges of Light and Compact EV Inverter*

Figure 8.4 depicts the breakdown of a Si inverter with liquid cooling. It has more than ten different heterogeneous components. Moreover, over half of the inverter's weight and volume is accounted for by the heat sink and capacitor. The inverter's cooling rails and liquid cycling system can be taken out by adopting an air cooling technique. Additionally, the power module, dc-link capacitor, and heat sink in the inverter can be decreased by adopting SiC devices in place of their Si counterparts and benefiting from the increased switching frequency.

Some obstacles need to be removed in order to actualize an air-cooling SiC inverter for next-generation EV inverters.

1. **Barrier 1**: How to develop a systemic design methodology from the standpoint of air cooling to combine the many parts of the inverter.
2. **Barrier 2**: How to multi-physics optimize the important parts of the power module, dc-link capacitor, and thermal management.

8.3 The Proposed Design Methodology for Air-Cooling Inverter

Every component of the air-cooling SiC inverter should be used to its best potential in order to achieve compact design and heterogeneous integration goals. Therefore, adequate attention should be given to the inverter's optimal design. The inverter's co-design technique is proposed and is depicted in Fig. 8.5.

Electric-thermal–mechanical co-design is used in Fig. 8.5 from a multi-physics approach to improving the packaging structure and material of the power module. The minimal capacitance of the dc-link capacitor is optimized with regard to the

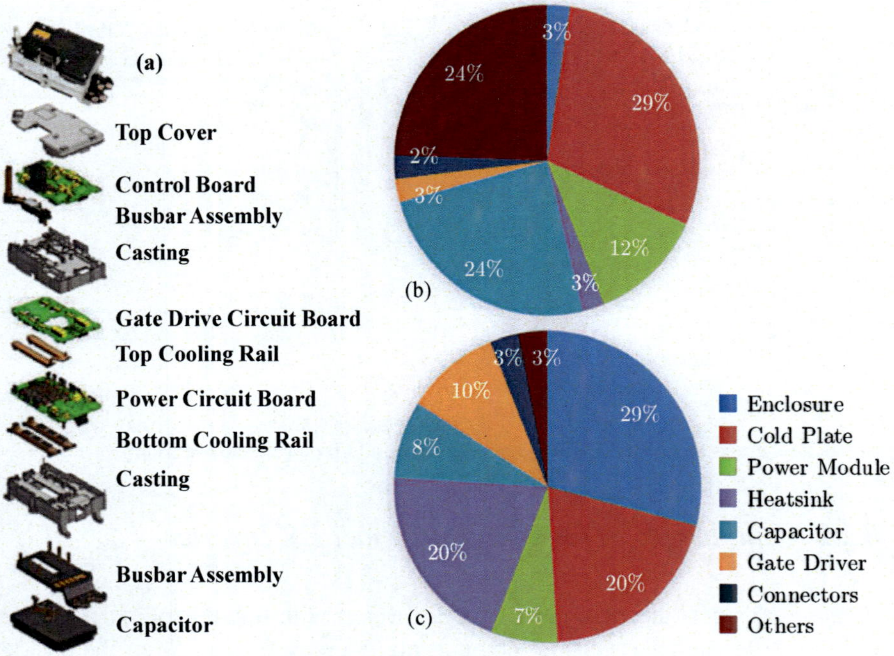

Fig. 8.4 Teardown of liquid-cooling EV inverter. **a** Teardown of the inverter, **b** volume breakdown of the inverter, and **c** weight breakdown of the inverter

specific switching frequency f_s, modulation ratio m, power factor $\cos\theta$, and ripple voltage Δv_{dc}. In terms of thermal management, electro-thermal co-design is proposed in order to optimize the heat sink with an emphasis on the tradeoff between thermal resistance, cost, and weight.

8.4 The Proposed Design Methodology for Air-Cooling Inverter

In this Section, the ideal design processes for the power module, dc-link capacitor, and heat sink are provided step-by-step in accordance with the co-design methodology in Fig. 8.5.

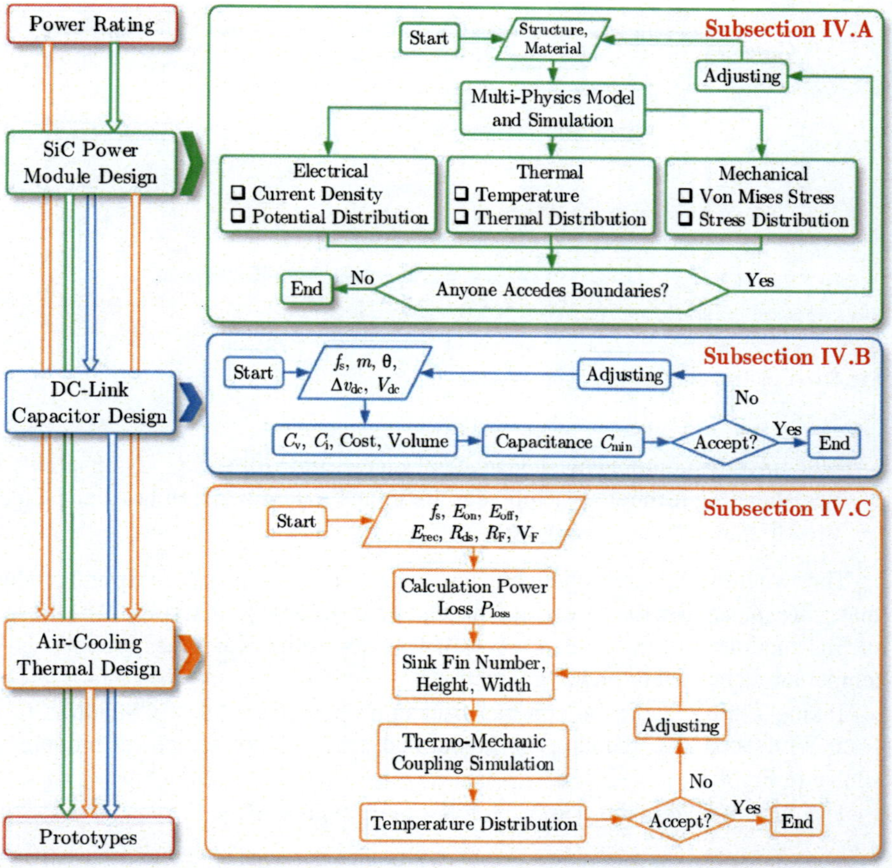

Fig. 8.5 Co-design methodology of air-cooling inverter in power module, dc-link capacitor, and heat sink levels

8.4.1 *The Proposed Multi-physics-Based Design of SiC Power Module*

This subsection presents the modeling and simulation of the power module utilizing a multi-physics method, as seen in Fig. 8.5. In order to improve the design of the power module prototype, some observed principles are put into practice.

1. Multi-physics-oriented modelling of the power module

   Figure 8.6 shows how a wire-bonding power module is put together. High-temperature solder is used to link the chip to the direct-bonded copper (DBC) layer. The sandwich-shaped DBC is constructed of copper, ceramic, and copper. The principal obstacle to the power module's thermal resistance is typically a ceramic substance with low heat conductivity, such as Al_2O_3, AlN, or Si_3N_4. As a result, the ceramic layer's height is constrained, making the DBC vulnerable. The baseplate is used to support the DBC and lessen stress on the power module and

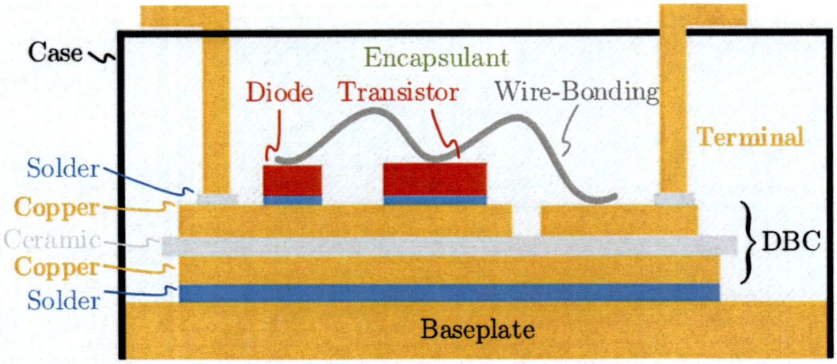

Fig. 8.6 Configuration of wire-bonding power module

other inverter components. It is made of CuW, CuMo, or AlSiC. Using ultrasonic wire-bonding technology, chip pads are joined to terminals. Both Al and Cu are used to make the bonding wire.

The switching loss and conducting loss of the SiC device heat the chip and produce material expansion in the power module due to the electro-thermal stress. The lifespan of SiC modules decreases as a result of thermo-mechanical stress brought on by temperature changes or power fluctuations.

Taking a 50 A/1.2 kV six-in-one packaging power module CCS050M12CM2 from Wolfspeed into account, the layout and schematic of the power module are shown in Fig. 8.7.

The current field in the power module can be expressed as

$$\begin{cases} \boldsymbol{U} = -\nabla \boldsymbol{\varphi} \\ \boldsymbol{J} = \gamma \boldsymbol{U} \end{cases} \tag{8.1}$$

where the intensity of the electric field $\boldsymbol{U}$ stands for the potential voltage $\boldsymbol{\varphi}$ gradient. The current density is $\boldsymbol{J}$. γ is the material's electrical conductivity. It is possible to

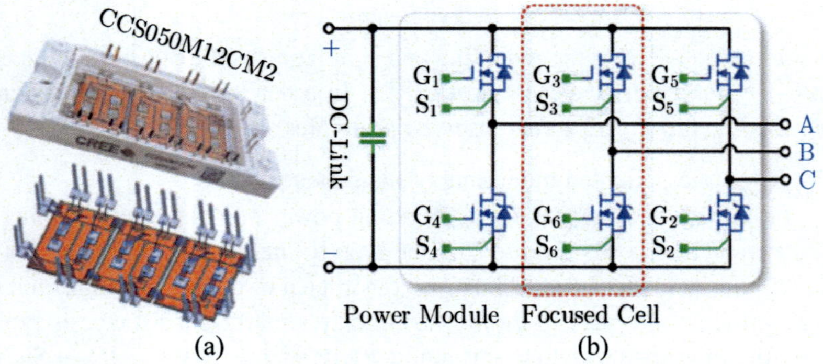

Fig. 8.7 Configuration of a six-in-one power module. **a** Inner layout and **b** equivalent circuit

express the gradient of $\boldsymbol{J}$ as

$$\nabla \boldsymbol{J} = \nabla \gamma(-\nabla \boldsymbol{\varphi}) = \boldsymbol{Q}_{\mathrm{j}} \tag{8.2}$$

where $\boldsymbol{Q}_{\mathrm{j}}$ is boundary current sources.

The power losses of conductors and chips are the heat sources in the power module. According to the current field, the power loss per volume Q_{v}, in W/m^3, can be written as

$$Q_{\mathrm{v}} = |\boldsymbol{J}|^2/\gamma \tag{8.3}$$

Temperature is a factor that affects electrical conductivity, and it can be stated as

$$\begin{cases} \gamma_{\mathrm{SiC}} = a_{\mathrm{SiC}} T + \gamma_{\mathrm{SiC0}} \\ \gamma_{\mathrm{Metal}} = \frac{\gamma_{\mathrm{Metal0}}}{1 + a_{\mathrm{Metal}}(T - T_{\mathrm{ref}})} \end{cases} \tag{8.4}$$

where γ_{SiC0} and γ_{Metal0}, in S/m, are electrical conductivities of SiC and metal materials at the reference temperature T_{ref}. a_{SiC} and a_{Metal} are temperature-dependent coefficients. T is the material's actual working temperature. As can be seen, the electrical conductivity of the SiC material rises with temperature, but the metal material exhibits the opposite behavior.

Fourier's rule states that the thermal conduction in the power module can be represented as follows when taking into account the heat source Q_{v}:

$$\nabla(k \nabla T) + Q_{\mathrm{v}} = \rho c \frac{\partial T}{\partial t} \tag{8.5}$$

k is the heat conductivity, measured in W/(m·°C). In J/(kg·°C), c denotes thermal specific. ρ denotes the substance's density, expressed in kg/m^3. For the regions of the power module without a heat source, such as the ceramic and baseplate, (8.5) can be expressed more simply as

$$\nabla(k \nabla T) = \rho c \frac{\partial T}{\partial t} \tag{8.6}$$

The power module's mixed multi-layer structure causes mechanical stresses in the module as a result of the mismatched thermal expansion coefficients of the various layers. There are two components to the overall strain tension. Temperature causes the first portion of the equation, ε^{T}, while stress causes the second component, ε^{E}, which can be represented as

$$\begin{cases} \varepsilon = \varepsilon^{\mathrm{T}} + \varepsilon^{\mathrm{E}} \\ \varepsilon^{\mathrm{T}} = \xi(T - T_{\mathrm{ref}}) \\ \varepsilon^{\mathrm{E}} = 0.5\left[\nabla u + (\nabla u)^{\mathrm{T}}\right] \end{cases} \tag{8.7}$$

u being the displacement The coefficient of thermal expansion is ξ. There are several tensor equations that can be used to describe the mechanical stress distribution in the power module.

$$\begin{cases} \frac{\partial \sigma_{ij}}{\partial x_j} + f_i = \rho \frac{\partial^2 u_i}{\partial t^2} + \mu \rho \frac{\partial u_i}{\partial t} \\ \varepsilon_{ij} = \frac{1}{2}\left(\frac{\partial u_i}{\partial x_j} + \frac{\partial u_j}{\partial x_i}\right) = \varepsilon_{ij}^{\mathrm{T}} + \varepsilon_{ij}^{\mathrm{E}} \\ \varepsilon_{ij}^{\mathrm{E}} = \sigma_{ij}/D_{ijkl} \\ \varepsilon_{ij}^{\mathrm{T}} = \xi \Delta T \delta_{ij} \end{cases} \tag{8.8}$$

where σ_{ij} is the stress tensor. f_i is external stress. μ is the damping coefficient. x_1, x_2, and x_3 represent x-, y-, and z-axis, respectively. D_{ijkl}, $i, j, k, l \in \{1, 2, 3\}$, is the tensor of elastic modulus according to the generalized Hooke's law, which can be determined by

$$D_{ijkl} = \frac{E}{1+\upsilon}\delta_{ik}\delta_{jl} + \frac{E}{(1+\upsilon)(1-2\upsilon)}\delta_{ij}\delta_{kl}\upsilon \tag{8.9}$$

E stands for Young's modulus. The poisson ratio is denoted by the letter ν. The Dirac function-δ has the following notation:

$$\delta_{ij} = \begin{cases} 1 \; i = j \\ 0 \; i \neq j \end{cases} \tag{8.10}$$

It can be deduced from (8.7) and (8.8) that

$$\begin{cases} \varepsilon_{ij} = \varepsilon_{ij}^{\mathrm{E}} + \xi \Delta T \delta_{ij} \\ \varepsilon_{ij}^{\mathrm{E}} = \frac{1}{2}\left(\frac{\partial u_i}{\partial x_j} + \frac{\partial u_j}{\partial x_i}\right) - \xi \Delta T \delta_{ij} \end{cases} \tag{8.11}$$

The σ_{ij} in (8.8) can be rewritten as

$$\sigma_{ij} = 0.5 D_{ijkl}(u_{kl} + u_{lk}) - \xi \nabla T D_{ijkl}\delta_{kl} \tag{8.12}$$

In conclusion, Fig. 8.8 illustrates the coupling between the electrical, thermal, and mechanical stresses in the power module. According to (8.1)–(8.12), a finite element analysis (FEA) tool can be used to calculate the electro-thermal–mechanical stress distribution based on the boundary conditions.

2. Multi-physics-based simulation of the power module

   In the prepared six-in-one power module prototype, bare dies of SiC MOSFET H1M120N060 and SiC SBD H2S120N060 from Hestia are employed. Summarized datasheets of bare dies are indicated in Table 8.1.

Al_2O_3 is used as the ceramic material for DBC in the developed power module. The baseplate is made of copper. Materials $Sn_{96}Ag_{3.5}Cu_{0.5}$ and $Sn_{60}Pb_{40}$ are used

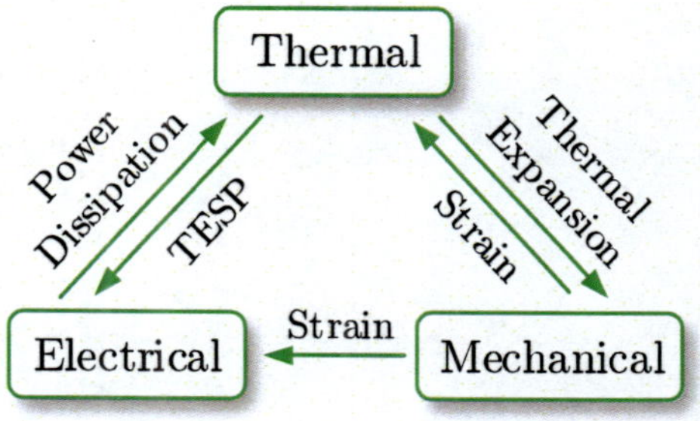

Fig. 8.8 Coupling principles of multi-physics in power module

Table 8.1 Specifications of SiC chips

Device	Voltage rating (kV)	Current rating	Size (μm)	Height (μm)
SiC MOSFET	1.2	41 A @ 25 °C	4290 × 2916	350
SiC SBD	1.2	35 A @ 132 °C	4250 × 4250	370

Table 8.2 Properties of materials in SiC power module

	SiC	$SnAg_{3.5}Cu_{0.5}$	$Sn_{60}Pb_{40}$	Al_2O_3	Cu	Al
γ (MS/m)	–	8.33	5.88	–	58.14	35.34
ρ (kg/m^3)	3210	7300	8400	3780	8960	2700
k [W/(m K)]	490	35	50	15	380	238
c [J/(kg K)]	800	226	167	30	390	900
ξ (10^{-6}/K)	4.4	23	25	6.5	17	23
v	0.142	0.347	0.4	0.22	0.37	0.33
E (GPa)	410	40	30	370	110	70

as solders for die attachment and DBC attachment, respectively. Heavy metal wire bonding has a 580 μm radius. Table 8.2 provides a list of the material's qualities.

An FEA-based simulation study is carried out to design the SiC power module. The steady-state stress distributions in the multi-physics are shown in Fig. 8.9.

According to Fig. 8.9a, the parasitic voltage drop on wire bonding causes power loss on the wire. In addition, a high temperature is present where the wire and chip make contact, and a thermally stretched wire causes additional stress at the wire's foot point, which could indicate a weakness in the wire-bonding power module. Therefore, materials with high electrical conductivity are useful for enhancing wire dependability. In order to lower thermomechanical stress and parasitic resistance and inductance, it is also advantageous to employ more parallel wires.

The current density at corners in the DBC is significantly higher than that in other areas, as seen in Fig. 8.9b. In addition, the wires are where the current density is concentrated. L-type traces and thin bonding wire should be avoided in order to lower current density because areas with higher current densities can also have more severe EMI problems.

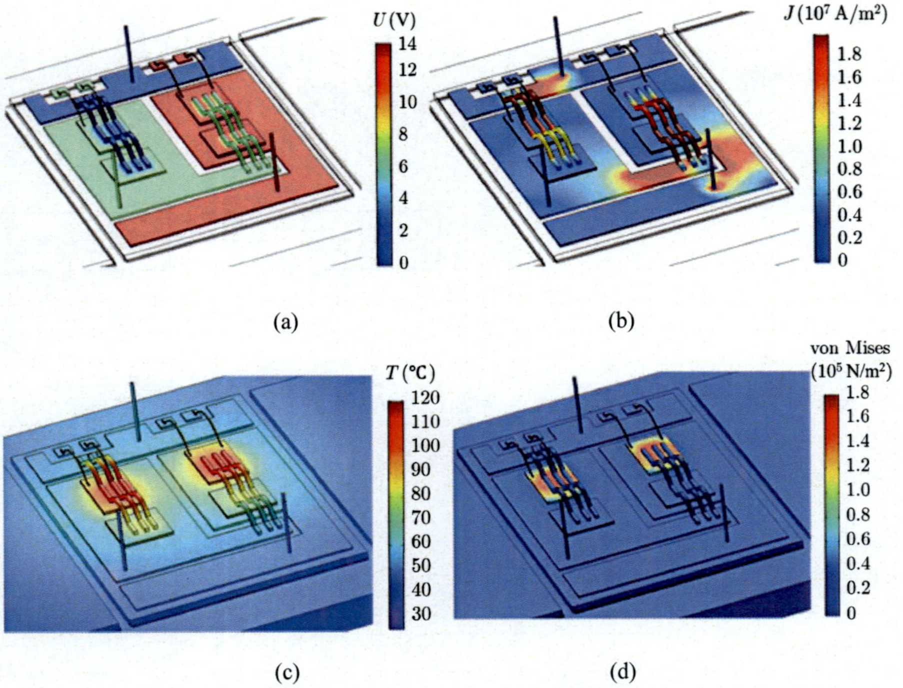

Fig. 8.9 Simulation results of six-in-one power module in **a** potential voltage, **b** current density, **c** temperature, and **d** von Mises stress

The chip should have enough space away from the edge of the DBC, as shown in Fig. 8.9c, taking the DBC layout into account, in order to lower junction-case thermal resistance and junction temperature. The local overheating can be minimized by the > 5 mm edge margin and inhibited thermal coupling.

The terminal is also a weak place that can fail, as shown in Fig. 8.9d, where the maximum mechanical stress at the foot point is 6.90×10^8 N/m^2. 5.59×10^8 N/m^2 is the maximum tension at the wire. For a realistic EV inverter, the bond wire is simple to lift off during power cycling or temperature cycling. At the corner of the die attachment, there is a 4.59×10^8 N/m^2 maximum stress around the chip. Typically, a power module's die attachments, bond wires, and terminals need to be strengthened.

The power loop should take precautions based on the multi-physics simulation to reduce current density and EMI problems. Additionally, copper should be used to make the terminals in order to lower the current density. In light of these concerns, an enhanced six-in-one power module is created, as depicted in Fig. 8.10.

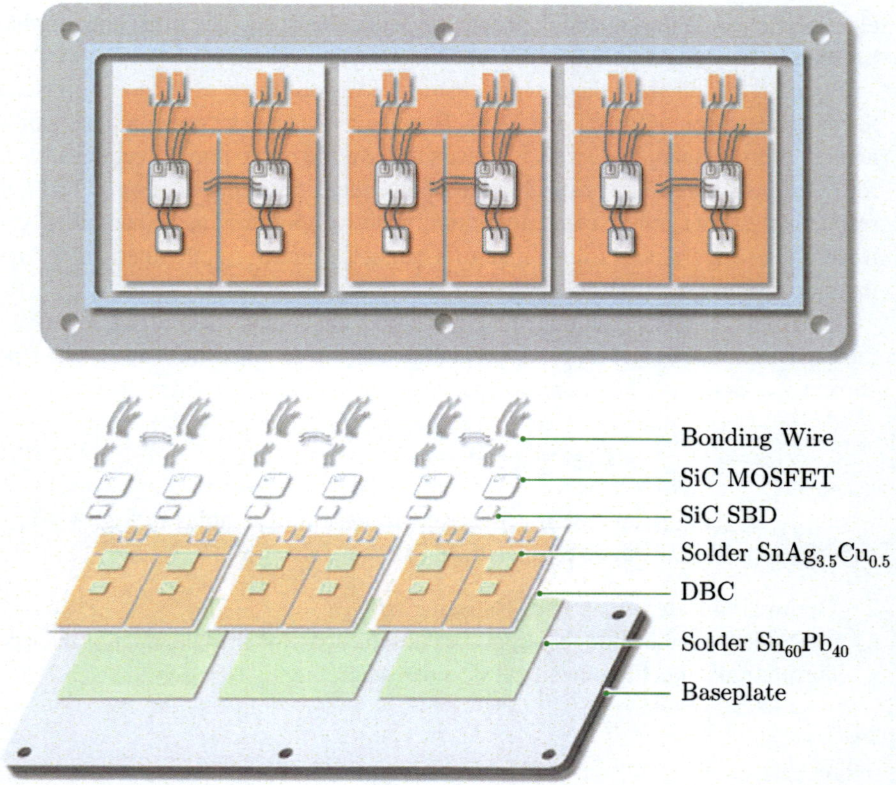

Fig. 8.10 Concept design of an enhanced six-in-one power module

8.4.2 The Proposed Optimal Selection of DC-Link Capacitance

As seen in Fig. 8.5, the capacitance and volume of the dc-link capacitor should be tuned to meet light and compact requirements. The capacitor is also one of the weakest parts of the inverter because of ripple voltage and ripples current. The chosen material has a significant impact on the capacitor's ability to handle ripple voltage and ripple current. The capacitor experiences ripples as a result of the SiC MOSFET switching, which reduces the capacitor's lifespan and causes power loss and excessive temperature.

1. Specification modelling of candidate capacitors

There are three types of capacitors that are frequently used for power converters that use various types of dielectrics: ceramic, electrolytic, and film capacitors. Although the ceramic capacitor can withstand high temperatures, it has a tiny capacitance and is humidity-sensitive. The electrolytic capacitor, on the other hand, cannot withstand high temperatures due to its huge capacitance; under these conditions, the electrolytic would dry up and the capacitor will be harmed. The film capacitor has a lifespan in-between that of ceramic and electrolytic capacitors.

Figure 8.11a displays the power densities of various types of capacitors based on the datasheets from KEMET and TDK. The electrolytic capacitor, which has a power density of more than 4 μF/cm^3, is proven to have the maximum power density. The film capacitor and ceramic capacitor have power densities of 1 μF/cm^3 and 2 μF/cm^3, respectively. The ceramic capacitor is significantly more expensive than other types, it should be mentioned. Film capacitors might be an affordable alternative for an inverter application.

The ripple current capacities of the capacitors are compared in Fig. 8.11 based on the capacitors in Fig. 8.11a, b. Modeling the capacitors' maximum allowable ripple current as a quadratic function of capacitance, which can be written as

$$I_{\mathrm{crms}} = f(C) = a_1C^2 + a_2C + a_3 \tag{8.13}$$

where a_1, a_2, and a_3 are coefficients. According to the samples in Fig. 8.11b, the parameters in (8.13) can be estimated, as listed in Table 8.3.

2. Optimizing Determination of Capacitance
   Aside from the voltage rating, the chosen capacitor's ripple voltage and ripple current must meet the inverter's dc-link requirement.

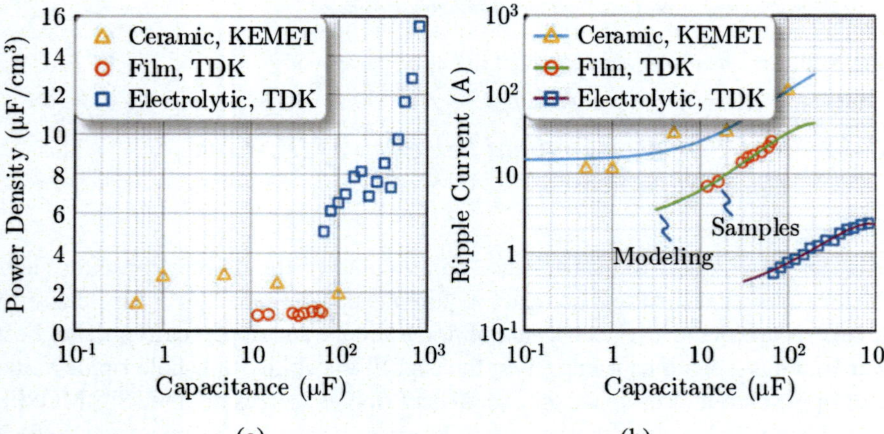

Fig. 8.11 Characteristics of capacitors. **a** Power density and **b** ripple current

Table 8.3 Parameters of capacitor ripple current models

Capacitor	a_1	a_2	a_3
Electrolytic capacitor	-2.86×10^{-6}	4.92×10^{-3}	0.28
Film capacitor	-1.00×10^{-3}	0.41	2.25
Ceramic capacitor	-1.32×10^{-3}	1.10	14.98

When ripple voltage is taken into consideration, the minimal capacitance C_v can be written as

$$C_v = \frac{\sqrt{6}}{2}\frac{I_{ac}}{f_s \Delta v_{dc}}\left[\frac{\sqrt{3}}{2} - \frac{3}{4}m\sin\frac{\pi}{3}\right]m\cos\theta \tag{8.14}$$

I_{ac} stands for the maximum load current per phase. The switching frequency is f_s. The ripple voltage's amplitude is given by Δv_{dc}, modulation ratio is $m = \sqrt{2}V_{ac}/V_{dc}$. Line-line RMS output voltage and dc-link input voltage are denoted by V_{ac} and V_{dc}, respectively. The angle of the power factor is θ. It goes without saying that increasing the switching frequency can lower capacitance. Additionally, modest dc-link capacitance is likewise influenced by increasing V_{dc} or decreasing m.

Regarding the ripple voltage problem, Fig. 8.12 shows how ripple voltage, switching frequency, and modulation ratio affect the dc-minimum link's capacitor. It has been discovered that using a high switching frequency and low modulation ratio can significantly lower the dc-requirement link's for capacitance. Additionally, the appropriate dc-link capacitance can be conserved by utilizing a capacitor with high ripple voltage capability.

The capacitor must also be able to withstand the ripple current brought on by the switching of SiC devices. The capacitor's minimal ripple current capability, I_{pmin}, can be written as

$$I_{pmin} = I_{ac}\sqrt{2m\left[\frac{\sqrt{3}}{4\pi} + \cos^2\theta\left(\frac{\sqrt{3}}{\pi} - \frac{9}{16}m\right)\right]} \tag{8.15}$$

The load current, modulation ratio, and power factor all affect ripple current. I_{ac}, m, and $\cos\theta$ are constants for a particular inverter. According to (8.13) and (8.15),

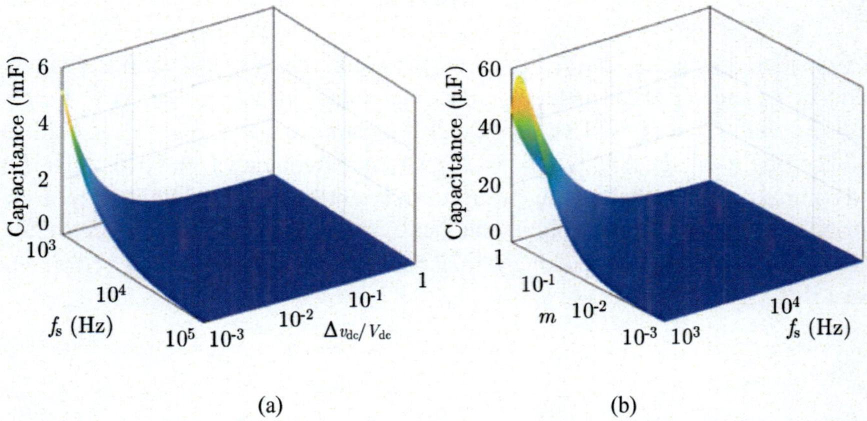

Fig. 8.12 Ripple voltage based minmum dc-link capacitor affected by **a** switching frequency and ripple voltage ratio, **b** switching frequency and modulation ratio

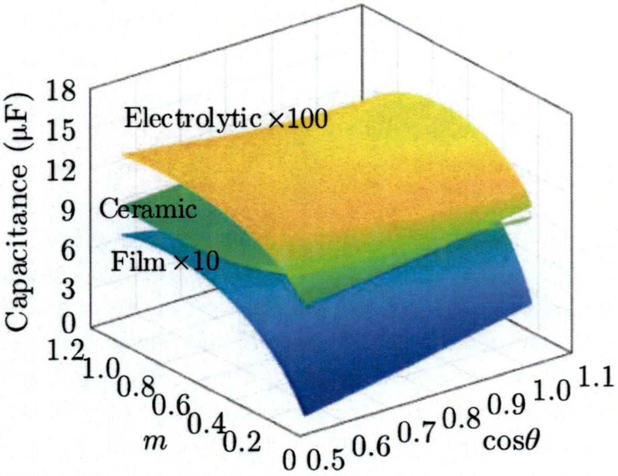

Fig. 8.13 Ripple current determined minimum dc-link capacitor affected by power factor and modulation ratio

the minimal capacitance can be produced as follows to support the ripple current

$$C_{\mathrm{i}} = f^{-1}(I_{\mathrm{pmin}}) = \frac{\sqrt{a_2^2 - 4a_1(a_3 - I_{\mathrm{pmin}})} - a_2}{2a_1} \tag{8.16}$$

Using various capacitor materials, Fig. 8.13 shows the desired minimum capacitor for the ripple current. The minimum capacitance of a dc-link can be reduced by more than 90% when ceramic and film capacitors are used in place of electrolytic capacitors. The excellent ripple current capability of film and ceramic capacitors is remarkably helpful to reduce the complexity of the dc-link.

To select the capacitance of dc-link C_{dc}, the ripple voltage and ripple current should be simultaneously satisfied, which can be expressed as

$$C_{\mathrm{dc}} = \max(C_{\mathrm{v}}, C_{\mathrm{i}}) \tag{8.17}$$

Considering various Δv_{dc}, Fig. 8.14 illustrates the C_{v} for various switching frequencies and capacitor materials under the conditions of $V_{\mathrm{dc}} = 600$ V, $m = 0.8$, $\cos\theta = 1$, and $I_{\mathrm{ac}} = 20$ A. The ideal candidate capacitors are presented in Table 8.4 with the ripple voltage $\Delta v_{\mathrm{dc}}/V_{\mathrm{dc}} = 5\%$ at switching frequency $f_{\mathrm{s}} = 50$ kHz. The light and compact performances of the inverter can be enhanced by switching to a ceramic capacitor instead of an electrolytic one, but the price goes up. The film capacitor is chosen in this chapter after taking into account the trade-off between performance and cost.

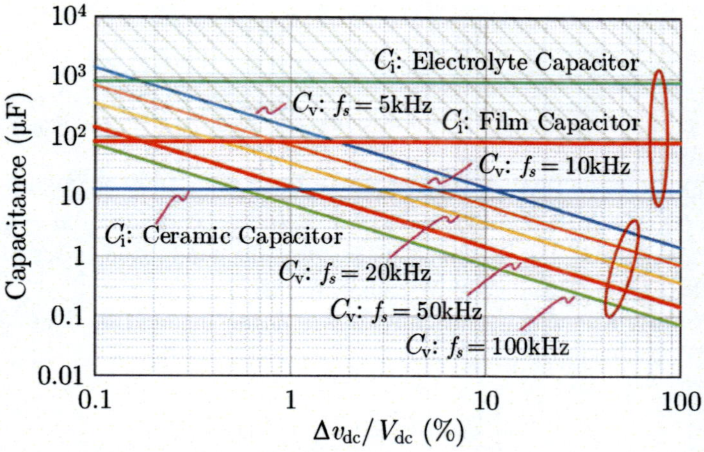

Fig. 8.14 Optimum selection of dc-link capacitance

Table 8.4 Optimal DC-link capacitors of different materials

Capacitor	Items	Capacitance (μF)	Weight (g)	Volume (cm^3)	Cost
Electrolytic	6 parallel, 2 series 330 μF, 450 V, B43504A5337M	900	624	423.9	\$92
Film	4 parallel 30 μF, 800 V, B32776G8306K	120	200	226.8	\$94
Ceramic	120 parallel 0.1 μF, 1 kV, C1812V104KDRACTU	12	3.24	2.94	\$195

8.4.3 The Proposed Thermal Design of Heat Sink

In this subsection, the power losses of the SiC devices are modeled, as seen in Fig. 8.5. The heat sink for air cooling is thoroughly optimized taking into account various fin structures, heat transfer coefficients, and materials.

1. Power Losses of SiC Devices

The output power of the SiC inverter feeding to the load is

$$P_o = \sqrt{3} V_{ac} I_{ac} / \sqrt{2} \tag{8.18}$$

where V_{ac} is the inverter's output line-to-line RMS voltage. The conduction losses of the SiC MOSFET and SiC diode, $P_{Mc,}$ and P_{Dc}, for a three-phase inverter with bipolar sine pulse width modulation (SPWM), are stated as

$$\begin{cases} P_{\text{Mc}} = \left(\frac{1}{8} + \frac{m\cos\theta}{3\pi}\right) R_{\text{ds,on}} I_{\text{ac}}^2 \\ P_{\text{Dc}} = \left(\frac{1}{8} - \frac{m\cos\theta}{3\pi}\right) R_{\text{F}} I_{\text{ac}}^2 + V_{\text{F0}} I_{\text{ac}} \end{cases} \tag{8.19}$$

where the MOSFET and diode's on-resistances are $R_{\text{ds,on}}$, and R_{F}, respectively. The diode's forward voltage is V_{F0}.

For the SiC MOSFET H1M120N060 and SiC diode H2S120N060 from Hestia, $R_{\text{ds,on}} = 60\ \text{m}\Omega$, $R_{\text{F}} = 35\ \text{m}\Omega$, $V_{\text{F0}} = 1.6$ V. In the condition of $m = 0.8$, $\cos\theta = 0.886$, $P_{\text{o}} = 10$ kW, $V_{\text{ac}} = 380$ V, $I_{\text{ac}} = 20$ A, the conduction losses of MOSFET and diode can be yielded as 5.5 W and 35 W, respectively.

In a similar manner, the switching losses of the P_{Ms} and P_{Ds}, SiC diodes, and MOSFETs can be represented as

$$\begin{cases} P_{\text{Ms}} = \frac{1}{\pi} f_s (E_{\text{on}} + E_{\text{off}}) \frac{V_{\text{dc}} I_{\text{m}}}{V_{\text{n}} V_{\text{n}}} \\ P_{\text{Ds}} = \frac{1}{\pi} f_s E_{\text{rec}} \frac{V_{\text{dc}} I_{\text{m}}}{V_{\text{n}} I_{\text{n}}} \end{cases} \tag{8.20}$$

where V_{n} and I_{n} represent the devices' rated voltage and current. E_{on} and E_{off} stand for the SiC MOSFET's turn-on and turn-off energy losses under the conditions of nominal voltage and current. E_{rec} is the SiC diode's reverse recovery loss under the specified test conditions. By applying an inductor-clamped double-pulse test and the turn-on and turn-off trajectories of SiC devices, these switching losses can be calculated as follows:

$$\begin{cases} E_{\text{on}} = \int_0^{t_{\text{on}}} v_{\text{ds}} i_{\text{d}} \text{d}t \\ E_{\text{off}} = \int_0^{t_{\text{off}}} v_{\text{ds}} i_{\text{d}} \text{d}t \\ E_{\text{rec}} = \int_0^{t_{\text{rec}}} v_{\text{F}} i_{\text{F}} \text{d}t \end{cases}, \tag{8.21}$$

where the times for the turn-on, turn-off, and reverse recovery procedures for SiC devices are t_{on}, t_{off}, and t_{rec}. As a result, the average power losses of each SiC MOSFET and SiC diode throughout each line-frequency period can be calculated as

$$\begin{cases} P_{\text{M}} = P_{\text{Mc}} + P_{\text{Ms}} \\ P_{\text{D}} = P_{\text{Dc}} + P_{\text{Ds}} \end{cases} \tag{8.22}$$

The total power loss for the six-in-one power module is

$$P_{\text{all}} = 6(P_{\text{M}} + P_{\text{D}}) \tag{8.23}$$

According to the datasheets of devices provided by Hestia, it is shown that $E_{\text{on}} = 115\ \mu\text{J}$, $E_{\text{off}} = 165\ \mu\text{J}$, $E_{\text{rec}} = 10\ \mu\text{J}$ in the conditions of $V_{\text{n}} = 800$ V and $I_{\text{n}} = 20$ A. In conditions of $P_{\text{o}} = 10$ kW, $V_{\text{dc}} = 600$ V, and $f_{\text{s}} = 50$ kHz, according to (8.20)–(8.22), it can be derived that $P_{\text{Ms}} = 3.6$ W and $P_{\text{Ds}} = 0.5$ W.

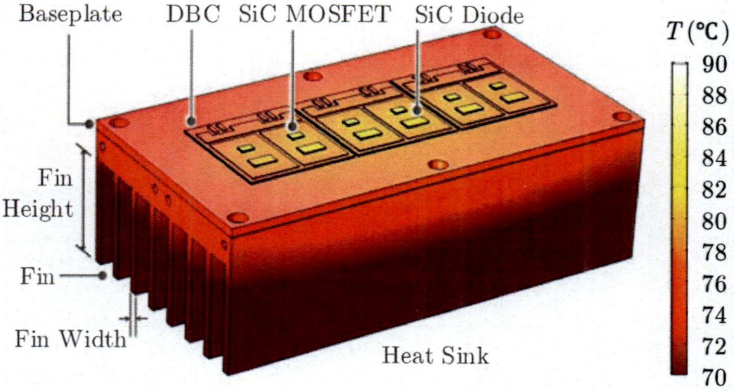

Fig. 8.15 Thermal distribution of the SiC inverter

2. Electro-Thermal Simulation of SiC Inverter
   An FEA method is used to ensure the SiC inverter's capacity for heat dissipation utilizing the multi-physics analysis tool COMSOL. The dc-link voltage of 600 V, load current of 20 A, and switching frequency of 50 kHz are the operational parameters. To simulate forced air cooling, the ambient temperature is 25 °C, and the heat sink's heat transfer coefficient is 50 W/(m^2·K). $P_M = 5.5 + 3.6 = 9.1$ W and $P_D = 35 + 0.5 = 35.5$ W respectively, estimate the power losses for each MOSFET and diode prospectively.

The chip's highest junction temperature is 83 °C, as shown by the simulation result in Fig. 8.15. The power dissipation of SiC devices can be satisfied by the thermal design.

Using various fin numbers, fin widths, fin heights, and materials, scanning results for various chip sizes are shown in Fig. 8.16. Large chip sizes are beneficial in lowering the junction-ambient thermal resistance R_{th} of the SiC inverter.

The results of the multi-physics simulation shown in Fig. 8.16a show that the thermal resistance does not visibly decrease when the number of fins increases past eight. Similar to this, as shown in Fig. 8.16b, the thermal resistance almost changes when the fin width is more than 4 mm. As illustrated in Fig. 8.16, when the heat sink's fin height exceeds 50 mm, the thermal resistance almost changes (c). In addition, Fig. 8.16 shows how the heat sink's materials affect thermal resistance (d). When the thermal conductivity exceeds 200 W/(m·K), it has been noticed that the thermal resistance almost completely changes. Aluminum is an economical option for the heat sink when the cost of the material is taken into account. Manufacturing is more complex when the fin number is higher. Meanwhile, a heavier heat sink results from bigger fins. As a result, there are compromises between thermal resistance, price, and weight. In order to accommodate this, an aluminum heat sink with eight pins, fins that are 4 mm wide and 50 mm tall, and junction-ambient thermal resistance of 0.36 K/W was constructed.

The heat transfer coefficient h_c of forced air cooling is determined by the air speed v_a, which can be, in experience, expressed as

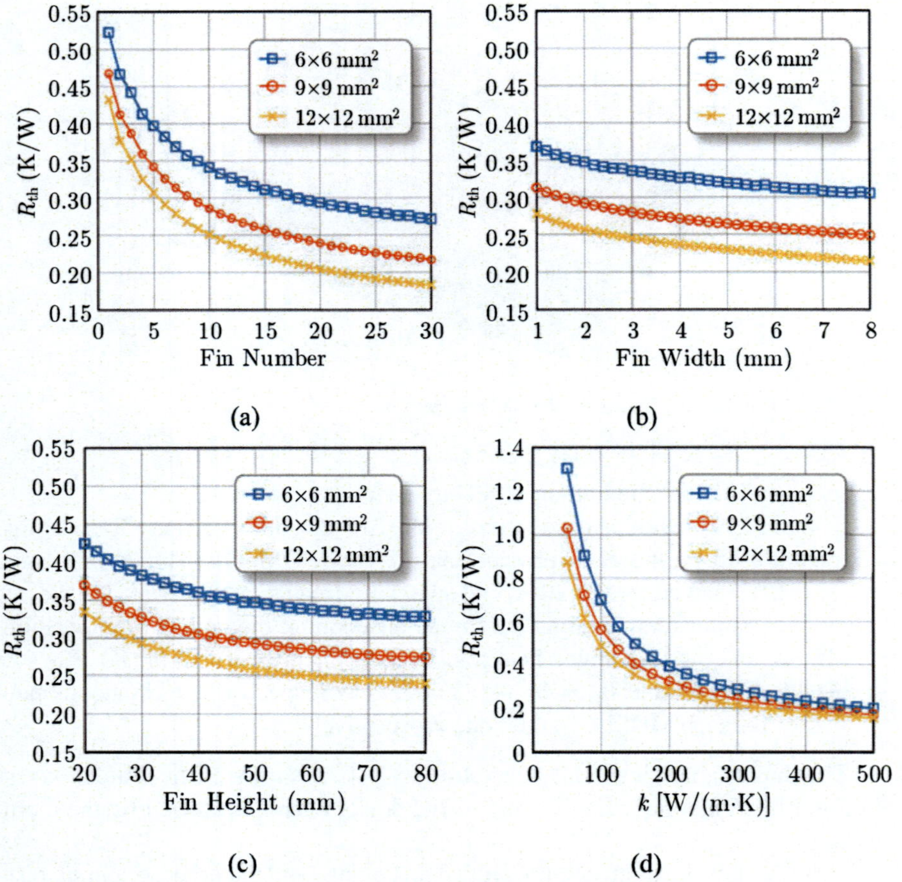

Fig. 8.16 Junction-ambient resistance of the heat sink. **a** Aluminum heat sink by using different fin numbers in condition of fin width 2 mm and fin height 50 mm, and **b** aluminum heat sink by using different fin width in condition of fin width number 8 and fin height 50 mm, **c** aluminum heat sink by using different fin height in condition of fin width number 8 and fin width 2 mm, and **d** heat sink influenced by thermal conductivity of material

$$h_c = 18.3 v_a^{0.6} \tag{8.24}$$

In Fig. 8.17a, the quantified coefficient is shown. It is clear that a big v_a helps to lower the heat resistance and increase the coefficient h_c. Large v_a, on the other hand, refers to bulk fans with a large volume, high cost, and high loss. As shown in Fig. 8.17b, the thermal resistance of the heat sink does not visibly decrease when the h_c surpasses 30 W/(m^2·K), and the matching wind speed provided by the fan is 3 m/s.

In this chapter, two fans SUNON 6015 (60 × 60 × 15 mm^3, 42 g) for inverters with wind speeds more than 3 m/s, wind fluxes of 18 cfm, and rotational speeds of 3600 rpm are gathered.

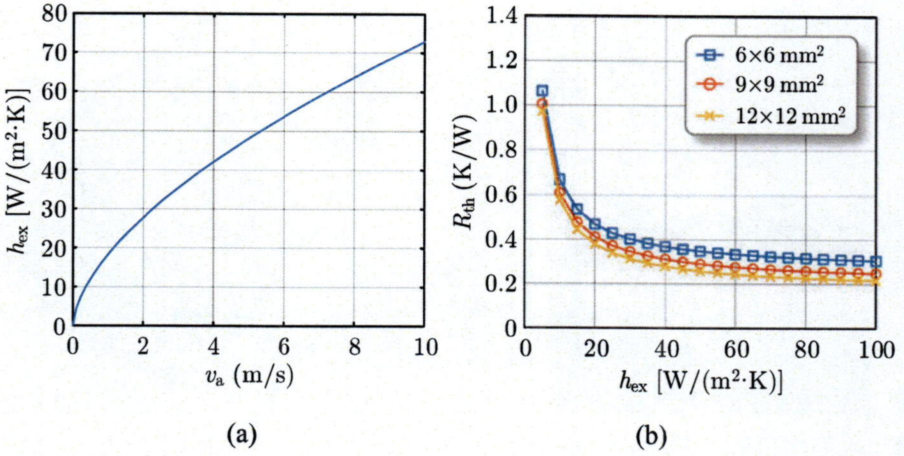

Fig. 8.17 Thermal resistance of the heat sink affected by heat transfer coefficient. **a** h_c versus v_a and **b** thermal resistance versus h_c

8.5 Experimental Results

The six-in-one SiC power module that was built and manufactured is exhibited in this Section to attest to the viability of the proposed design process. Additionally, the SiC inverter with air cooling has additional heterogeneous integration. Additionally, this Section proposes the experimental findings of the constructed power module and air-cooling inverter prototypes.

8.5.1 Prototypes of Power Module and Air-Cooling Inverter

A 50 A/1.2 kV six-in-one wire-bonding power module is created for a 10 kW three-phase inverter in accordance with the previously indicated multi-physics-oriented design, as shown in Fig. 8.18a.

Based on Sect. 8.3's ideal design. A SiC inverter is incorporated on the basis of the manufactured SiC power module, as shown in Fig. 8.18b. According to the optimum selection of capacitance, four 30 μF/800 V film capacitors B32776E8306K from TDK are used to form a 120 μF dc-link.

The inverter has a total volume of 0.8 L. The inverter has a power density of 12.5 kW/L. Figure 8.19a shows the inverter's volume breakdown. The inverter weighs 1.5 kg overall and has a 6.7 kW/kg power density. The weight distribution is shown in Fig. 8.19b. The inverter costs $ 485 in total. Figure 8.19c shows the expense breakdown. Figure 8.19 shows that the heat sink accounts for more than 60% of the inverter's volume and weight; this should be further optimized. In addition, the power module accounts for roughly 80% of the inverter's cost because SiC chips are

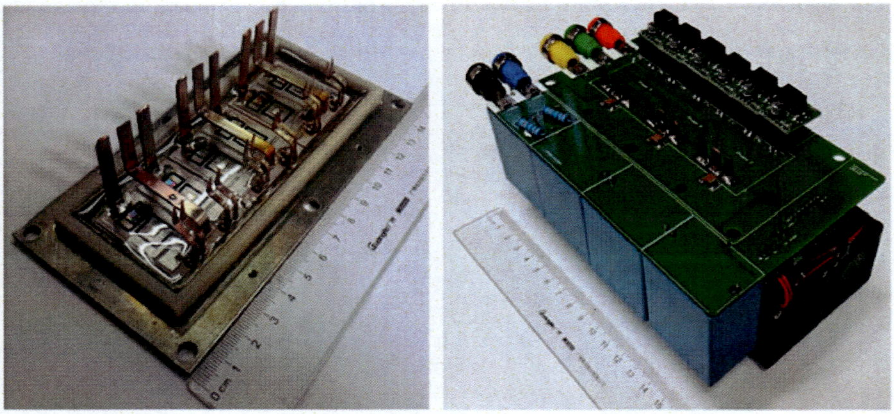

(a) (b)

Fig. 8.18 Prototypes of **a** fabricated SiC power module and **b** integrated air-cooling inverter

so expensive; however, this cost is predicted to reduce as SiC chip prices continue to fall.

A test bench is set up, as shown in Fig. 8.20, to verify the viability of the packaged SiC power module and heterogeneously integrated air-cooling inverter. The prototypes are controlled by a TI TMS 320F28335 DSP control board. The extremely quick transient voltage and current waveforms are recorded using A 1 GHz Digital oscilloscope 610Zi, a 200 MHz differential voltage probe DP6150B, and a 20 MHz flexible Rogowski coil CP9006S.

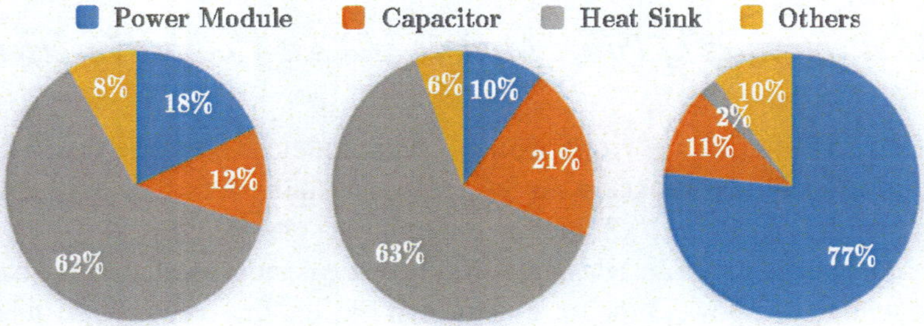

(a) (b) (c)

Fig. 8.19 Breakdown of SiC inverter by **a** volume, **b** weight, and **c** cost

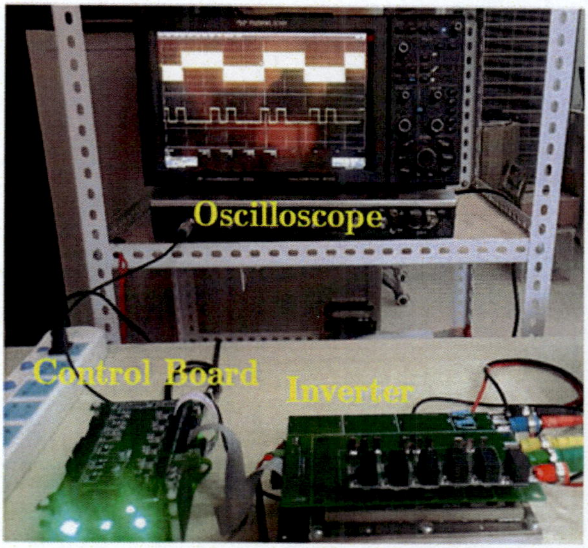

Fig. 8.20 Experimental platform of the prototypes

8.5.2 Experimental Results of SiC Power Module

In the condition of junction temperature 25 °C, dc-link voltage $V_{dc} = 600$ V, and gate resistance 10 Ω, Fig. 8.21a shows the measured switching trajectories of SiC MOSFET in the condition of different load currents from 10 to 40 A. As seen, the SiC device performs a very fast switching speed, and the switching time is within 50 ns. Meanwhile, the switching time increases with the load current.

Figure 8.21 displays the experimental turn-on-off trajectories at dc-link voltage 600 V and load current 40 A under various junction temperature conditions ranging from 25 to 150 °C (b). As can be shown, the junction temperature has a significant impact on the SiC MOSFET's turn-on and turn-off procedures.

In addition, switching losses associated with junction temperature are determined under various load current conditions, as illustrated in Fig. 8.22. By raising the junction temperature, it is possible to lower the threshold voltage, accelerate switching, and cut turn-on losses. However, the junction temperature or the load current has the greatest impact on the turn-off loss. The turn-on loss is also influenced by junction temperature and load current since it contains the diode's delicate reverse recovery loss, however, the relative deviation of temperature-dependent turn-on loss is less than 20%.

8.5.3 Experimental Results of SiC Inverter

The dc-link voltage for the air-cooled SiC inverter is 600 V. Line-line RMS output voltage is 220 V, and the modulation ratio is 0.9. 50 kHz is the switching frequency.

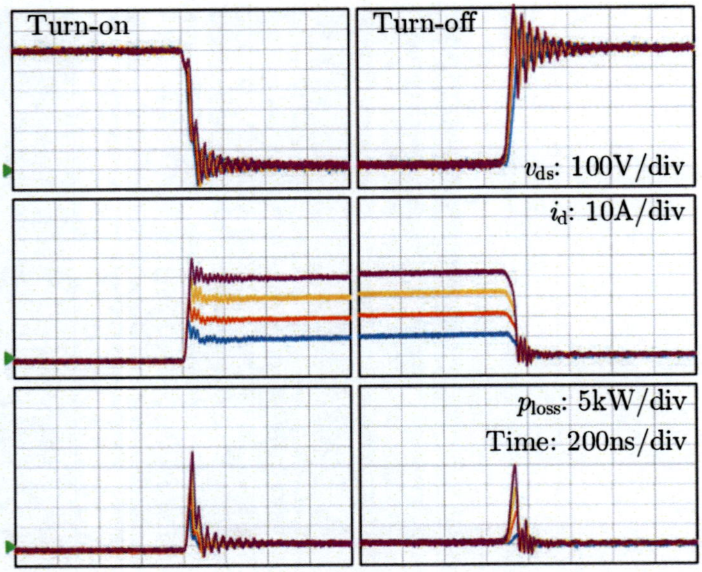

(a)

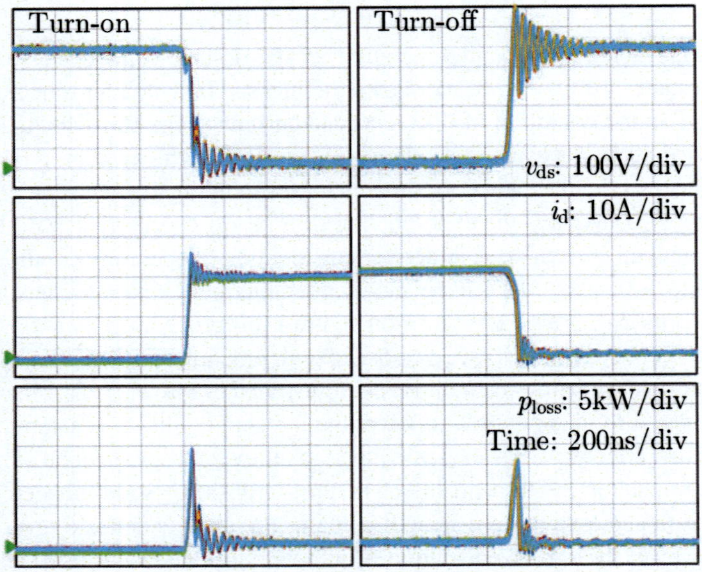

(b)

Fig. 8.21 Experimental turn on-off trajectories of the fabricated power module in the condition of **a** different load currents and **b** different junction temperatures

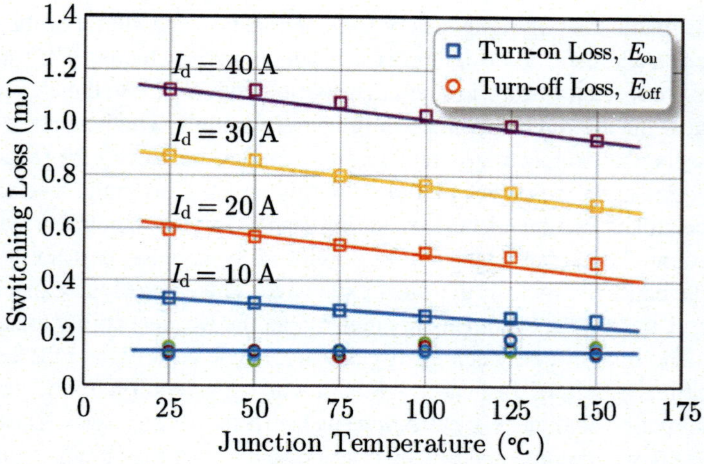

Fig. 8.22 Experimental swtiching losses in the condition of different load currents and junction temperatures

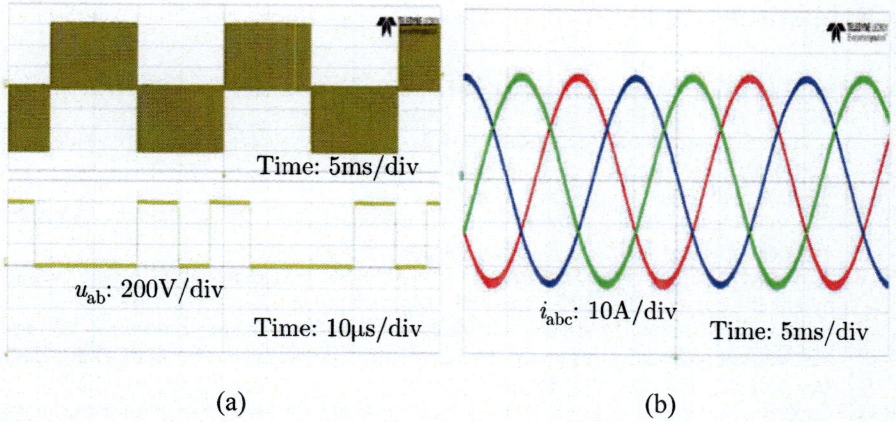

Fig. 8.23 Experimental results of the inverter. **a** The line-line voltage at legs and **b** phase current at loads

The motor is simulated by an RL load with $R = 10\ \Omega$ and $L = 1$ mH. Figure 8.23 shows the line-line voltage and phase current experimental findings.

8.6 Conclusions

The air-cooling inverter carrying on the SiC device can primarily improve the light and compact aims of the powertrain for EV deployment by removing the auxiliary liquid cooling system. The design approach and heterogeneous integration are still problems in this kind of highly customized application situation. For the air-cooling SiC inverter, a holistic co-design process is put out in this chapter. The multi-physics model is proposed to show the stress distribution with regard to the electro-thermal–mechanical interaction mechanism in the power module. The best power traces die attachments, and connecting wires are discovered to be crucial for reducing multi-physics stresses in the power module utilizing an FEA program. The minimal dc-link capacitor is computed best by taking ripple voltage, ripple current, and cost into account. The ceramic capacitor is believed to have the best ability to handle ripple current, but it is also the most expensive. The electrolytic capacitor, on the other hand, has the maximum energy density but the weakest ability to handle ripple current. The trade-off between energy density, ripple current, and the cost is better with the film capacitor. The heat sink is built with consideration for the size, material, and construction challenges by utilizing an electro-thermal analysis. The customized six-in-one power module and SiC inverter prototypes are then created and put to the test in order to validate the viability of the suggested design approach.

References

1. K. Rajashekara, "Present status and future trends in electric vehicle propulsion technologies," *IEEE J. of Emerg. and Sel. Topics Power Electron.*, vol. 1, no. 1, pp. 3–10, 2013.
2. S. Jahdi, O. Alatise, C. Fisher, L. Ran, and P. Mawby, "An evaluation of silicon carbide unipolar technologies for electric vehicle drive-trains," *IEEE J. of Emerg. and Sel. Topics in Power Electron.*, vol. 2, no. 3, pp. 517–528, 2014.
3. Z. J. Shen and I. Omura, "Power semiconductor devices for hybrid, electric, and fuel cell vehicles," *Proc. of the IEEE*, vol. 95, no. 4, pp. 778–789, 2007.
4. H. Zhang, L. M. Tolbert, and B. Ozpineci, "Impact of SiC devices on hybrid electric and plug-in hybrid electric vehicles," *IEEE Trans. Ind. Appl.*, vol. 47, no. 2, pp. 912–921, 2011.
5. K. Hamada, M. Nagao, M. Ajioka, and F. Kawai, "SiC—Emerging power device technology for next-generation electrically powered environmentally friendly vehicles," *IEEE Trans. Electron Dev.*, vol. 62, no. 2, pp. 278–285, 2015.
6. F. Wang, Z. Zhang, T. Ericsen, R. Raju, R. Burgos, and D. Boroyevich, "Advances in power conversion and drives for shipboard systems," *Proc. of the IEEE,* vol. 103, no. 12, pp. 2285–2311, 2015.
7. P. G. Neudeck, R. S. Okojie, and L. Y. Chen, "High-temperature electronics - a role for wide bandgap semiconductors?" *Proc. of the IEEE*, vol. 90, no. 6, pp. 1065–1076, 2002.
8. D. Peftitsis, R. Baburske, J. Rabkowski, J. Lutz, G. Tolstoy, and H. P. Nee, "Challenges regarding parallel connection of SiC JFETs," *IEEE Trans. Power Electron.*, vol. 28, no. 3, pp. 1449–1463, 2013.
9. P. Ning, F. Wang, and D. Zhang, "A high density 250°C junction temperature SiC power module development," *IEEE J. of Emerg. and Sel. Topics in Power Electron.*, vol. 2, no. 3, pp. 415–424, 2014.
10. Z. Chen, Y. Yao, D. Boroyevich, K. D. T. Ngo, P. Mattavelli, and K. Rajashekara, "A 1200-V, 60-A SiC MOSFET multichip phase-leg module for high-temperature, high-frequency applications," *IEEE Trans. Power Electron.*, vol. 29, no. 5, pp. 2307–2320, 2014.
11. P. Ning, F. Wang, and K. D.T. Ngo, "Automatic layout design for power module," *IEEE Trans. Power Electron.*, vol. 28, no. 1, pp. 481–487, 2013.
12. H. Li, S. M. Nielsen, S. Bęczkowski, and X. Wang, "A novel DBC layout for current imbalance mitigation in SiC MOSFET multichip power modules," *IEEE Trans. Power Electron.*, vol. 31, no. 12, pp. 8042–8045, 2016.

13. Z. Zeng, W. Shao, H. Chen, B. Hu, W. Chen, H. Li, and L. Ran, "Changes and challenges of photovoltaic inverter with silicon carbide device," *Renew. Sust. Energy Rev.*, vol. 78, pp. 624–639, 2017.
14. M. Wang, F. Luo, and L. Xu, "A double-end sourced wire-bonded multichip SiC MOSFET power module with improved dynamic current sharing," *IEEE J. of Emerg. and Sel. Topics in Power Electron.*, vol. 5, no. 4, pp. 1828–1836, 2017.
15. Z. Liang, P. Ning, and F. Wang, "Development of advanced all-SiC power modules," *IEEE Trans. Power Electron.*, vol. 29, no. 5, pp. 2289–2295, 2014.
16. C. Chen, Z. Huang, L. Chen, Y. Tan, Y. Kang, and F. Luo, "A flexible PCB based 3D integrated sic half-bridge power module with three-sided cooling using ultra-low inductive hybrid packaging structure," *IEEE Trans. Power Electron.*, vol. PP, no. 99, pp. 1–1, 2018. (Early Access)
17. S. Seal, M. D. Glover, and H. A. Mantooth, "3-D wire bondless switching cell using flip-chip-bonded silicon carbide power devices," *IEEE Trans. Power Electron.*, vol. 33, no. 10, pp. 8553–8564, 2018.
18. C. Yao, Z. Wang, W. Li, H. Li, J. Qian, C. Han, F. Luo, and Jin Wang, "Comparison study of common-mode noise and thermal performance for lateral wire-bonded and vertically integrated high power diode modules," *IEEE Trans. Power Electron.*, vol. 33, no. 12, pp. 10572–10582, 2018.
19. A. Dutta and S. S. Ang, "A module-level spring-interconnected stack power module," *IEEE Trans. Comp., Pack. and Man. Tech.*, vol. 9, no. 1, pp. 88–95, 2019.
20. H. Wang and F. Blaabjerg, "Reliability of capacitors for DC-link applications in power electronic converters—An overview," *IEEE Trans. on Ind. Appl.*, vol. 50, no. 5, pp. 3569–3578, 2014.
21. J. Falck, C. Felgemacher, A. Rojko, M. Liserre, and P. Zacharias, "Reliability of power electronic systems: An industry perspective," *IEEE Ind. Electron. Mag.*, vol. 12, no. 2, pp. 24–35, 2018.
22. H. Wen, W. Xiao, X. Wen, and P. Armstrong, "Analysis and evaluation of DC-link capacitors for high-power-density electric vehicle drive systems," *IEEE Trans. Veh. Tech.*, vol. 61, no. 7, pp. 2950–2964, 2012.
23. B. Karanayil, V. G. Agelidis, and J. Pou, "Performance evaluation of three-phase grid-connected photovoltaic inverters using electrolytic or polypropylene film capacitors", *IEEE Trans. Sust. Energy*, vol. 5, no. 4, pp. 1297–1306, 2014.
24. A. Mallik, A. Khaligh, "Intermediate DC-link capacitor reduction in a two-stage cascaded AC/DC converter for more electric aircrafts", *IEEE Trans. Veh. Tech.*, vol. 67, no. 2, pp. 935–947, 2018.
25. M. Mellincovsky, V. Yuhimenko, M. M. Peretz, and A. Kuperman, "Analysis and control of direct voltage regulated active DC-link capacitance reduction circuit," *IEEE Trans. Power Electron.*, vol. 33, no. 7, pp. 6318–6332, 2018.
26. H. Wang, Y. Liu, and H. Wang, "On the practical design of a two-terminal active capacitor," *IEEE Trans. on Power Electron.*, vol. PP, no. 9, pp. 1–1, 2019.
27. X. Cao, Q. C. Zhong, and W. L. Ming, "Ripple eliminator to smooth DC-bus voltage and reduce the total capacitance required," *IEEE Trans. Ind. Electron.*, vol. 62, no. 4, pp. 2224–2235, 2015.
28. M. Chen, K. K. Afridi, C. Sombuddha, and D. J. Perreault, "Multitrack power conversion architecture," *IEEE Trans. Power Electron.*, vol. 32, no. 1, pp. 325–340, 2017.
29. P. Ning, X. Wen, L. Li, "Interpretation method investigation of converter system layout design," in *Proc. of IEEE IECON*, 2017, pp. 4949–4954.
30. I. Laird, X. Yuan, J. Scoltock, and A. J. Forsyth, "A design optimization tool for maximizing the power density of 3-phase DC–AC converters using silicon carbide (SiC) devices," *IEEE Trans. Power Electron.*, vol. 33, no. 4, pp. 2913–2932, 2018.
31. Y. Ko, M. Andresen, G. Buticchi, and M. Liserre, "Discontinuous-modulation-based active thermal control of power electronic modules in wind farms," *IEEE Trans. Power Electron.*, vol. 34, no. 1, pp. 301–310, 2019.

32. T. Wu, Z. Wang, B. Ozpineci, M. Chinthavali, and S. L. Campbell, " Automated heatsink optimization for air-cooling power semiconductor modules," *IEEE Trans. Power Electron.*, vol. PP, no. 99, pp. 1–1, 2018. (Early Access)
33. D. Christen, M. Stojadinovic, and J. Biela, "Energy efficient heat sink design: Natural versus forced convection cooling," *IEEE Trans. Power Electron.*, vol. 32, no. 11, pp. 8693–8704, 2017.
34. E. M. Dede, S. N. Joshi, and F. Zhou, "Topology optimization, additive layer manufacturing, and experimental testing of an air-cooling heat sink," *J. of Mech. Des.*, vol. 137, no. 11, pp. 1–9, 2015.
35. B. Hu, Z. Zeng, W. Shao, Q. Ma, H. Ren, H. Li, L. Ran, and Z. Li, "Novel cooling technology to reduce thermal impedance and thermomechanical stress for SiC application," in *Proc. of IEEE APEC*, 2017, pp. 3063–3067.
36. A. Lajunen, Y. Yang, and A. Emadi, "Recent developments in thermal management of electrified powertrains," *IEEE Trans. Veh. Tech.*, vol. 67, no. 12, pp. 11486–11499, 2018.
37. M. A. Ebadian and C. X. Lin, "A review of high-heat-flux heat removal technologies," *J. Heat Transfer*, vol. 133, no. 11, pp. 1–11, 2011.
38. K. C. Otiaba, N. N. Ekere, R. S. Bhatti, S. Mallik, M. O. Alam, and E. H. Amalu, "Thermal interface materials for automotive electronic control unit: Trends, technology and R&D challenges," *Microelectronics Reliability*, vol. 51, no. 12, pp. 2031–2043, 2011.